One Road To Poultry Success

A Poultry Primer Intended For The Novice Collected During Many Years of Experience in the Poultry Business

by Pittsfield Poultry Farms

with an introduction by Jackson Chambers

This work contains material that was originally published in 1917.

This publication is within the Public Domain.

This edition is reprinted for educational purposes and in accordance with all applicable Federal Laws.

Introduction Copyright 2017 by Jackson Chambers

IMPORTANT NOTE & DISCLAIMER

IMPORTANT NOTE :

As with all reprinted books of this age that are intended to perfectly reproduce the original edition, considerable pains and effort had to be undertaken to correct fading and sometimes outright damage to existing proofs of this title.

A book can require total rebuilding of some pages from multiple different digital proofs. Despite this, imperfections still sometimes exist in the final proof and may detract slightly from the visual appearance of the text.

This book appears exactly as it did when it was first printed.

DISCLAIMER :

Due to the age of this book, some methods or practices may have been deemed unsafe or unacceptable in the interim years. In utilizing the information herein, you do so at your own risk.

We republish antiquarian books without judgment or revisionism, solely for their historical and cultural importance, and for educational purposes.

Please use due diligence before gifting an antiquated book, especially to a minor.

Self Reliance Books

Get more historic titles on animal and stock breeding, gardening and old fashioned skills by visiting us at:

http://selfreliancebooks.blogspot.com/

INTRODUCTION

I am very pleased to present to you another essential title for all novice poultry breeders and egg producers – *One Road to Poultry Success*. It was first published in 1917, one hundred years ago, and was written by *Pittsfield Poultry Farms Co.* in Holliston, Massachusetts, USA.

If you are embarking on a full-scale commercial operation, or if you are a back-yarder raising your own chickens for meat and eggs, this title will be absolutely priceless to you.

This title was written especially for the novice in the field, and is filled with expert knowledge of the crucial basics that make the difference between being a successful poultry operation, and one that falls by the wayside.

With chapters that cover everything from profits, location, choosing a breed, poultry housing, crops and more, this book will send you well on your way to successfully raising poultry.

Good luck, happy farming, and may your chickens never stop laying!

Jackson Chambers
State of Jefferson, December 2017

AN ACKNOWLEDGMENT

We desire to acknowledge our indebtedness to the Cyphers Incubator Company for suggestions obtained from the instructive and valuable bulletins issued by them. In some few instances sections of our text are verbatim extracts from the text of their bulletins, because the idea was so clearly and concisely expressed by them that it would detract from the value of the text to change the wording.

We especially appreciate the courtesy of the Cyphers Company in granting us permission to make this use of their bulletins. We recommend every poultryman, especially every beginner, who does not now have access to these bulletins, to make application to the Cyphers Company for them. We consider them among the most valuable contributions that have been made to poultry literature.

We also desire to acknowledge our indebtedness to our farm managers for their co-operation in supplying details, and to the Poultry Keeper Publishing Company for quotations from their book, "Poultry Keeping in a Nutshell."

THE PUBLISHERS

Table of Contents

CHAPTER ONE—Poultry Profits

A chapter which tells the truth about poultry profits and which explains why the hen always pays on the small farm and why it seldom pays adequately on the large farm devoted exclusively to poultry.

CHAPTER TWO—Location of Farm

Discussed from a standpoint of business and duty.

CHAPTER THREE—Choosing the Breed

Information regarding money-making characteristics of several popular breeds.

CHAPTER FOUR—Poultry Houses

Describing only those houses which have proved successful on the farms of the Pittsfield Poultry Farms Company.

CHAPTER FIVE—Feeding and Care of Chicks

Giving every detail of one successful way of caring for poultry — practically a reprint of a bulletin that has been distributed to the extent of nearly 100,000 copies.

CHAPTER SIX—Crops on the Farm

Given largely in the form of notes on account of the large amount of information to be crowded into a limited space.

CHAPTER SEVEN—Dual Farming

Devoted largely to the desirable combination of fruit and chicken farming.

CHAPTER EIGHT—Disposal of Poultry Products

How to obtain better prices and more profit.

CHAPTER NINE—General Topics

A brief discussion of a variety of subjects.

CHAPTER TEN—Fancy Breeding

Its pleasures and difficulties.

APPENDIX

Giving recipes and tables that are useful on the farm but which did not have any logical place in the text.

One Road to Poultry Success

PREFACE

In submitting the following chapters to our many friends, who, like ourselves, are looking for the surest and safest methods of insuring profitable returns in the poultry business, we do so with the hope that they will be received in the same spirit that prompted us to write them, that is, as a guide to the beginner, an incentive to him who has the making and means of success, and as a deterrent to him who has neither the ability nor the capital to avert disaster.

This modest pamphlet makes no pretence of being "most complete and comprehensive." It departs from the precedent usually followed by poultry writers. It deals with subjects which are always avoided — probably from fear of hurting the business — and it discusses subjects which many consider foreign to the poultry business; but if it is carefully studied by the beginner, we are sure that it will lead to most surprising and satisfactory results.

While we admit there are many equally good methods of operating the poultry business, we believe it far wiser to submit to the novice only one method — or, as the title of this pamphlet suggests, to point out from the many, only "**One Road to Poultry Success.**" This should be, and is, a method that is consistent from beginning to end and that we have thoroughly tried out in our own business experience. If several methods for handling each branch of the business are simultaneously put before the beginner, the chances are that he will select procedures in one branch that will not be consistent with procedures selected for other branches, which would lead to a net result of only confusion and bewilderment.

Owing to the fact that we are admittedly the world's largest producers of day-old chicks, it may appear to some that certain recommendations we make are based on prejudice. We have, however, leaned over backward, as the saying is, in trying to speak without prejudice. Surely we have no axe to grind in recommending proper houses and proper food, except that we wish our chicks to have the best chance possible to do well. In recommending certain breeds we may be to some extent prejudiced to the breeds we carry, but popular prejudice, as it happens, seems to run to the same breeds. Ask yourself, do we not carry every breed that is popular enough to be salable? The basis for popularity of course, is ability to produce. As a matter of fact we are personally in favor of Barred Rocks and Single Comb White Leghorns, but we are confident you will not discover it in the text.

To help pay the expenses of this book, we have sold advertising space to certain outside manufacturers and dealers. We have, however, limited the advertisements of these people to such goods as we have personally tested and used. Other goods may have just as much merit as the ones they call to your attention, but it is our judgment that there are none better; so unless the reader has had wide experience, we feel sure he will not go wrong in adopting the suggestions made herein by the different concerns, all of whom we know to be reliable.

CHAPTER ONE
POULTRY PROFITS

We wish to present this subject — Poultry Profits — to the reader in such a way that he can accurately gauge his chances of success, so that, if the venture looks feasible, he may look forward with intelligent optimism at a particular result to be obtained, or so that if on the other hand the future seems full of dangers and impossibilities, he may give up the project in time to save both his self-respect and his money.

We will not attempt to figure out the profits that should be made from each branch of the poultry business; such a task would require many times the space available. Nor will we attempt to give undisputed figures in any case, but we will attempt to point out some considerations that make for failure or success.

Some writers claim one should make $6.97 from each hen kept, or some other equally ridiculous figure. Then some other "wise ones" say that "no hen ever paid a cent." These latter try to substantiate their statement by the remark that hens are "either moulting or setting twelve months in the year" and offer no further argument except a knowing wink of the eye.

WHAT IS THE TRUTH AND WHY

It is no doubt true that there have been occasional small flocks of selected birds that have yielded profits as high as $6.97 per bird, or even higher, but such results are rare and much harm is done by giving them publicity, as they create a wrong impression in the mind of the novice considering the business. The only safe figures to follow are those that eliminate profits from especially favorable circumstances and that take into consideration only those circumstances that one can expect under every day conditions, figured down to a business basis in which all expenses are included.

To a certain degree, also, the "wise one" is speaking the truth when he says poultry doesn't pay, because he very probably has in mind some specific instance where the impossible has been attempted.

TO GET DOWN TO FACTS

To simplify matters in our discussion of the subject we will divide poultry keepers into two classes, viz. —

    1. That class that makes poultry a side line to some other business (it may be only one branch of a general farming business); the term commonly applied to this class is "Back Yard Poultry-keeping."

    2. That class that makes poultry the one and only business.

BACK YARD POULTRY-KEEPING

Working it as a side line, poultry-keeping does not necessarily increase overhead charges. Bills for rent, telephone, heat, light, etc., need not be charged to the poultry. We will show later that the capitalized value of these savings is approximately $4,000. Very frequently also such poultry-keeping, besides escaping the overhead charges, saves considerable on its labor expense, for there is much time during the year when regular help of other departments has little to do and may just as well be cleaning hen houses as loafing.

There are many waste products of other businesses also, such as sour milk, table scraps, green food, etc. that are convertible with great adaptability into food for hens. When poultry is a side line there can be no argument; a dozen, a hundred, five hundred, or any number of hens may, under any and all conditions, be made to pay a handsome profit.

Poultry as a side line of a fruit farm makes a particularly happy combination. The birds furnish splendid fertilizer for the trees, destroy millions of insects that ravage the fruit, keep the ground constantly cultivated, etc. in return for which the trees furnish the shade that is so necessary for the birds in hot weather.

A garden truck farm, also, is not complete without an auxiliary hen plant. No form of dressing has yet been devised that is so economical and resultful for growing vegetables as clear hen manure which hens produce in such liberal quantities. The practical value of this manure is as much as $20 per ton.

POULTRY AS A SOLE BUSINESS

Conditions are entirely different, however, when one turns to poultry-keeping as a sole means of earning a living. All overhead expenses, all labor, all mistakes of judgment, must, like every other expense, be charged directly to the poultry with the consequent effect on net results.

It may be taken as too obvious to need demonstration that a dozen hens cannot possibly yield an income that will pay the rent. On the other hand, figures from our experience will make it equally obvious that 5000 hens should pay all expenses with a good profit on top.

It is necessary to resort to actual figures, however, to intelligently arrive at any conclusion as to how much profit may be expected from any given number of hens, or in other words to learn just how many hens one must keep to be able to reasonably expect a profit.

In any plant an economical hen will take its share of food and from it produce products that are of greater value. The increase in value thus produced is the gross profit from that hen, and in the case of the "backyard plant" this gross profit is nearly all net profit. When the plant is run for poultry alone, this gross profit must first be applied to the items of expense that the "backyard plant" escapes, and if there are sufficient numbers of hens, the gross profit produced by them will be large enough to pay not only these items but also to leave a net profit.

The problem before us resolves itself into finding out how many hens will be required to produce sufficient gross profit to pay all the expenses besides the food and direct labor charges. When we know this, we shall know that any additional number of hens kept, will be producing a gross profit that is practically all net profit.

To state it in other words, when poultry is kept as a sole business there must be a certain number of hens to pay overhead charges; any additional number will produce the same profit they would on a "backyard plant."

The two items that use up the gross profits in the "poultry as a sole business" plant, and scarcely enter into the expense of the "backyard" poultry plant, are living expenses of the owner and interest on capitalized value of overhead expense. With such items as these in mind, it is our belief, and we shall try to prove it to you, that most people would fail to make both ends meet with even as many as 500 hens, unless, as previously stated, they are kept as a side line. And we also believe that real success does not come with much less than 1000 hens.

We are in the business of selling day-old chicks and breeding stock, so it seems like suicide for us to make such a statement, but we would prefer to lose our small profit than to encourage some one to lose his all, simply because he did not have sufficient capital to keep a flock large enough to be a profitable flock.

Having made the above outline of what we intend to prove, we will proceed to make the point clear by a few specific figures. We know that the results these figures show are being demonstrated every day, yet we do not pretend that they will fit all conditions. They will, however, serve to indicate a safe mode of procedure.

THREE CONDITIONS MAKE FOR SUCCESS

1. Adaptability — This includes proper location of farm, proper stock, proper houses, proper soil, personal health and ambition, knowledge of feeding, personal power of observation, etc. etc. As it is difficult to make any figures based on a personal equation, and as this book will give comprehensive suggestions on location of farm, feeding, etc., we will assume that all requirements are fulfilled in this respect.

2. Sufficient Capital or Credit.

3. Size — Enough layers to take care of all overhead expenses and enough additional layers to produce a substantial net profit.

Having eliminated the first condition we will discuss the other two. We will figure roughly on two concrete examples, viz, a 500 layer plant and a 2000 layer plant. Space will not permit us to figure into this particular part of the matter to any further extent. We have, however, made similar figures on plants of many other sizes and will give below a tabulation of the results in the form of curves.

ONE ROAD TO POULTRY SUCCESS

BASIS OF FIGURES

We will assume that it takes two years to get the proposition on to a good running basis, and that only half the houses and stock are provided for the first year; that there is no other business connected with it; that any broilers or eggs sold the first year go to pay miscellaneous expenses; that after the second year, the sale of broilers, two year old stock, etc., pays for the new stock required each year. In short, we leave it to the owner to find ways of getting extra profits over and above our figures. Our figures assume that the egg sales pay all expenses and are the only source of profit. This is the only safe way to figure.

FIRST YEAR'S EXPENSES IN ESTABLISHING A POULTRY PLANT

	Ultimate Size	
	500 Layers	2000 Layers
15 Acre Farm with Buildings	$2500.00	$2500.00
1000 Day-old Chicks	170.00	
4000 Day-old Chicks		680.00
2 Colony Brooder Houses Complete	100.00	
6 Colony Brooder Houses Complete		300.00
Additional 4 x 8 Colony Houses	48.00	192.00
Laying House	400.00	1250.00
Grain First Year	300.00	1200.00
Salary of Owner	900.00	900.00
Extra Help		100.00
Interest, Taxes, Insurance	290.00	420.00
Total	$4708.00	$7542.00

SECOND YEAR'S EXPENSES IN ESTABLISHING A POULTRY PLANT

Day-old Chicks	$170.00	$680.00
Laying House	400.00	1250.00
Grain for New Stock	300.00	1200.00
Grain for Year Old Stock	400.00	1600.00
Salary of Owner	900.00	900.00
Extra Help	50.00	140.00
Interest, Taxes, Insurance	425.00	775.00
Total Second Year	$2645.00	$6545.00
Total First Year	4708.00	7542.00
Two Year's Expenses	$7353.00	$14087.00
Less Eggs from year-olds	900.00	3600.00
Gross Capital	$6453.00	$10487.00
Mortgage	2000.00	3000.00
Net Capital	$4453.00	$7487.00

THIRD YEAR AND EACH AFTERWARDS

Owner's Salary	$900.00	$900.00
Feed of Old Birds	800.00	3200.00
Interest, etc.	425.00	775.00
Extra Labor	50.00	140.00
Expenses — Going Plant	$2175.00	$5015.00
Income — 10 Doz. Eggs at 30 cents	1500.00	6000.00
Loss on 500 Layers*	$675.00	
Profit on 2000 Layers*		$985.00

*Note—Owner's salary is included.

PROFIT AND LOSS CURVE ON BASIS OF NUMBER OF LAYERS

Not having room to show other detailed examples than those above we submit the chart which shows, in a general way, the profit or loss to be expected from any size plant, when poultry is kept as a sole business. Allowance should be made for the available price of eggs in your locality.

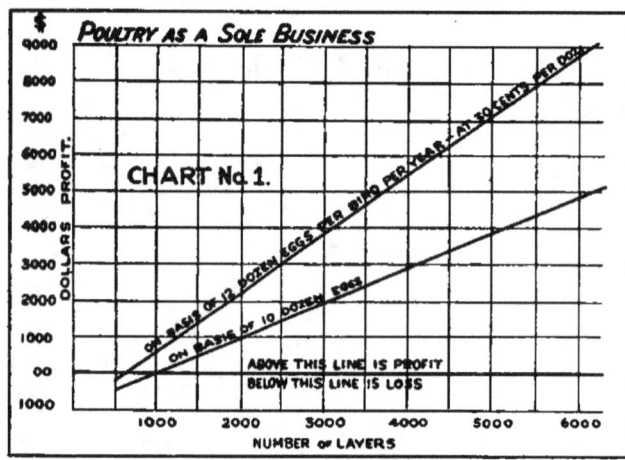

This chart shows clearly that 1000 layers is as small a plant as one can hope to operate as a sole business with any profit to himself above his salary. These figures should discourage anyone with small capital from going into the poultry business as a sole business. It is the lack of a thorough understanding of these figures that has been responsible for the poultry losses you have heard about. There is a way out of it and we propose to point out that "one road to success."

RESULTS APPLIED TO A BACK YARD PLANT

The profit one may reasonably expect from Back Yard Plants of various sizes is illustrated in Chart No. 2 in which the same figures are used as in Chart No. 1, except that the overhead expenses, that are otherwise taken care of, are eliminated. Assuming that our figures and bases of figuring are correct, this chart shows that poultry, as a side line, may be kept at a profit, regardless of the size of the plant or number of birds kept.

It shows further that the profit on each bird would be slightly larger than it would be on even the very largest plant operated for poultry alone.

The value of getting a large number of eggs from each bird is also shown: for example, there is over 400 per cent more profit from the hen that lays ten dozen eggs than there is from the hen that lays only eight dozen per year.

Capital Required

Chart No. 3 graphically illustrates the amount of capital required for a plant of any given size where poultry is the sole business. Where poultry is a side line the amount of capital required is obtained by deducting $4,000.

Overhead Expenses

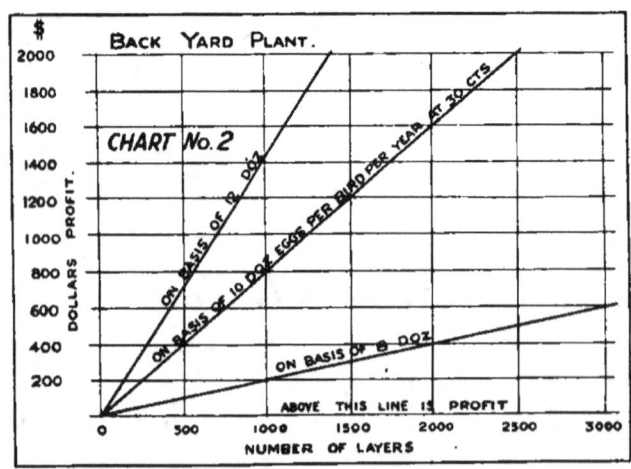

When one starts a poultry farm there are a few things that have to be done to about the same extent whether the ultimate capacity is to be 500 or 5000 layers. One cannot expect to buy a good farm with buildings on it for less than $2500. Of course if one intended to keep the larger number of layers, he would be willing to pay some more for his farm.

Besides the farm, one must spend about two years of his time before things are in shape to show a profit; we figure that this time is worth $900 a year, or a total of $1800. Adding this to cost of farm we have $4300.

CAPITAL REQUIRED WHERE IT IS SOLE BUSINESS

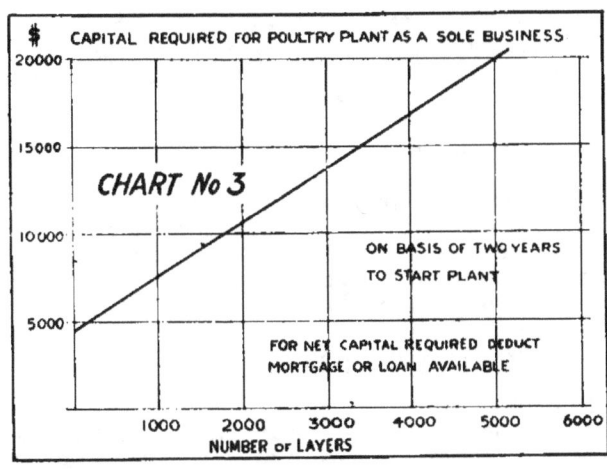

Chart No. 3 shows graphically that there is a capital outlay of $4300 before any layers or poultry equipment is installed. This expense is entirely eliminated in a back yard plant. It is easy to see that the greater number of layers there are, the less proportion of this basic expense will be charged against each layer. The hens do not, of course, have to pay the principal represented by this basic expense, but they do have to pay the interest. With the usual interest charges, the interest per year would amount to $240. There are other charges that should be added such as insurance, telephone, depreciation, etc. It would be setting it low to call the total of these charges $300 per year. If there were only 300 hens on the farm each hen would be charged $1 per year interest.

If there were 3000 hens each hen would be charged only 10 cents interest. It does not matter, however, how many layers a poultry plant may have: there would always be some charge against each hen which the back yard hens do not have to bear.

Chart No. 3 shows that a 5000 layer plant would cost approximately $20,000, and Chart No. 1 shows that one could expect a profit from this plant of from $4,000 to $7,000. In other words from 80 cents to $1.40 per hen. This is, of course, in addition to owner's salary, and to it can be added any extra profits from by-products.

A 2000 layer plant, as has been shown, can not do nearly as well. Such a plant should make between 50 cents to $1.00 per hen, the smaller amount on basis of 10 dozen eggs per hen and the larger amount on the basis of 12 dozen.

A Back Yard plant on the same basis would make between 85 cents and $1.45, regardless of how few or how many hens were kept.

All the above figures are on the basis of an average price of 30 cents per dozen for eggs.

AVERAGE PRICE OF EGGS

The average price received for eggs is dependent on two factors; viz, the distribution of production by months and the nature of the market for the eggs. From a careful census taken within a radius of twenty miles of Boston, we find that the Back Yard plants dispose of their eggs at an average price ranging from 32 to 36 cents per dozen, the average being for the year; that larger plants, sending some few eggs to Boston market, average as low as 30 cents, and where all eggs go to Boston, the average price drops to 29 cents. It would seem that 30 cents is a very conservative price to use as a basis, as anyone, by careful planning, should be able to find a market at least as good as that.

PROFIT AND LOSS PER YEAR ON BASIS OF EGG YIELD

The real key-note to the profits accruing from the operation of a plant, lies in the number of eggs obtained from the layers and the price obtained for the eggs. We append the following table which shows total net profit to be expected from plants of various sizes at varying yields per hen, eggs being figured at 30 cents per dozen.

Eggs per year per hen	Size of Plant. (As a sole business.)						Layers
	500	1000	2000	3000	4000	5000	
108			$400	$1100	$1800	$2500	
120			1000	2000	3000	4000	
132			1550	2900	4200	5500	
144		$600	2200	3800	5400	7000	
156		900	2800	4700	6700	8500	
168		1200	3400	5600	8000	10000	
180	$75	1500	4000	6500	9200	11500	
192	225	1800	4600	7400	10400	13000	

Where there are no figures, there would be a loss.

AN UNJUST REPUTATION

The poultry business has a reputation for failures that it does not deserve, when it is compared with other businesses. The fact has been conceded that 90 per cent that enter into business of any kind fail to make good. That is, taking into consideration every kind of business into which men enter, success is attained in not more than one instance out of ten.

NOT STRANGE

Knowing this to be true, can we consider it strange if we hear of a failure in the chicken business occasionally? There can be no doubt that there are many failures in this particular business, but we believe that the percentage of failures is much smaller than in many other lines of business. There are hundreds of thousands of people who raise poultry and we are of the opinion that a very large percentage of them do well.

STRIKING INSTANCE

We cannot help thinking that the reputation that the chicken business has for failures is acquired because of those striking instances where some man, entirely ignorant of the requirements of the poultry business, entering into it mostly as a fad, builds up a plant of tremendous proportions, with equipment and buildings provided regardless of investment costs. He employs a large crew of men, who receive high salaries, but who have no interest either in the business or in the employer except to draw down good, fat salaries as long as the fad lasts. Poultry business built upon such foundations sooner or later comes to grief, and the failure makes a big stir in the poultry world because of its magnitude.

NOT ALL FAILURES

Such ventures do not, however, represent any significant part of the poultry business, either in money invested or people interested. While we have no figures to demonstrate our statement, we believe that at least 90 per cent of the people engaged in poultry-keeping practice what we have called "Back Yard Poultry Keeping." The other ten per cent consist of those who conduct the business as a sole business. We are by no means willing to admit that the percentage of those who make a success of the business on the latter basis, is small. Not all of the ventures of this kind are failures.

Among those who conduct the business on the "Back Yard Poultry" basis, it is rarely that we hear of a failure. No statistics, of course, are available as to the success of those who keep poultry in this way, but we venture to say, without fear of contradiction, that a very large proportion of them are successful, and some of them to a very great degree.

NOT WITHOUT RISK

We do not wish it to be inferred that the risk of failure in the chicken business, on any basis, is remote. The chances for success or failure exist probably in this business to the same degree that they exist in most other businesses. Our object in writing at all on the matter is to try to offset the impression that prevails in the minds of many, that the risk of failure is exceedingly great.

POULTRY FOR PLEASURE

There are many people who raise poultry solely for the pleasure they derive from being among birds and from watching them develop. Very frequently it happens that this class of people do not make a success from a business standpoint, because they do not look at the venture from that angle. It is probable, however, that they find that poultry is as inexpensive a hobby as any could be. We imagine it is pretty near right to say that very few of such cases show a loss and very few much profit; that the majority of them just about pay expenses.

POULTRY AND EDUCATION

College education is not a necessary requisite to success in the poultry business. Such education, however, would doubtless lead one to make a somewhat better success of even this business. It is fair to assume that the greatest successes have been made by those people who are sufficiently intelligent to have been able to digest a college course, if the opportunity for one had been put in their way. In other words, anyone who has natural business aptitude should be easily successful with poultry from a profit-making standpoint, provided he has no pronounced dislike for birds or for some other feature of the business.

Conversely, poultry-keeping is not a vocation that it is beneath the educated man to consider. It offers reasonable opportunities for profit and leads to a life that is healthy and sane. More and more people of education are taking up the business each year, and when the time arrives, that the considerable bulk of the business is done by men of education and good business ability, the poultry business will become a very much more important factor in the business affairs of the nation than it is to-day.

CONCLUSIONS

From the above we believe the following conclusions may safely be arrived at:

That Back Yard Plants should make as much, or more, per hen as any large plant no matter how large the latter i.e., from 80 cents to $1.50 or more per hen per year.

That starting a poultry plant as a sole business is apt to be a failure, unless conditions are thoroughly understood.

That it takes over 1000 layers to assure a profitable poultry plant, where there is no other business.

That instead of the usual advise "GO SLOW" we would substitute:—

"Keep your poultry plant a Back Yard plant until you have nursed it up to a 1000 bird plant or have sufficient capital to make it such without crippling yourself." In other words, don't give up your job until you have your plant on a self-supporting basis.

20TH CENTURY HOUSE, SHOWING FRUIT CELLAR AND GARAGE UNDERNEATH.

CHAPTER TWO

LOCATION OF FARM

For the sake of continuity in our story, we will assume that the reader has definitely decided to enter the poultry business. The next matter, then, for him to consider will be the selection of a farm. It may very likely be true that he already owns a farm; in such case he should make every effort to adapt his farm to the business at hand. He should study what branch of farming would be most suited to his conditions to combine with the poultry. We take it for granted that some other line will be combined with the poultry, because we are convinced that the average novice will make a much greater success of his venture, if he takes up that variety of poultry-keeping that we have spoken of previously, as "Back Yard Poultry-keeping." We will not go into the advantages of such poultry-keeping here, because we have taken it up in more or less detail in the chapter entitled "Dual Farming," in which chapter we especially recommend fruit and poultry as the ideal combination.

It may be, however, that fruit would not be desirable for the conditions at hand. In which case, as we have said, some other line should be selected.

THE NEW FARM

Not knowing, of course, the circumstances connected with a farm already owned, we shall not attempt to say how such a farm can be best adapted to poultry. We shall take up the entire matter from the standpoint of one who is about to purchase an entirely new farm and desires to have it as near right as possible.

It seems to us that the first question to consider is the market that will be available for the poultry products. As a rule one should be so situated that he can make prompt deliveries to a city of considerable size. We do not mean by this that it is necessary to be located within ten, twenty, or thirty miles of such a city, but he should be so situated that he can make deliveries with few, if any, transfers, by express or parcel post, within say, twelve hours. Poultry products will lose none of their value in traveling that length of time.

There are certain peculiar locations in which one does not need a city for his market, as for instance, a location near a large hotel, or near a large hospital.

To be located in either of the above ways usually means that better prices will be obtained for the products, because they will arrive in strictly nice condition without delay, and because the poultry man will have a better opportunity to keep in touch with his market.

The advantages of this may be best illustrated by a specific, even though imaginary, instance. We will assume that a poultry man has one thousand layers, and that these layers each produce twelve dozen eggs per year. This would mean a total production of twelve thousand dozens of eggs to be disposed of. Assume further, that by being able to have a good market, he receives three cents per dozen more for his eggs than he would if located in some undesirable place. This means that he would have an extra profit of $360 per year. On this basis, figuring at a ten per cent rate, he could afford to have a farm in a good location, cost him $3600 more than one in a poor location.

IDEAL LOCATION

Of course, it would be highly satisfactory and pleasant to have the farm located on one of the main avenues of travel near the great cities, but the original cost of such farm is usually excessive and it requires a business of large proportions to support it.

The ideal location for the average reader will be one of the medium sized farms somewhat off the main boulevard near a big city. If the farm can be located at a reasonable expense on, or near, an electric car line, it will be found a distinct advantage. One should not go to the extreme of purchasing an isolated farm, where there is no travel passing, because doing so will mean the loss of a desirable business suggested below, and will mean less convenience and pleasure for his family.

MOTOR ROUTES

There is considerable advantage in being located close to a road on which automobiles frequently pass. Motorists are becoming quite generally habituated to purchasing their fresh eggs, poultry, vegetables, etc. direct from the farmer, and the farmer, who exercises a little ingenuity in advertising and offering his products to this class of trade, will derive a very considerable increased profit from it. And this trade will be found very willing to pay a good price for choice products and to be very appreciative of any effort made to please it.

CONSIDER THE FAMILY

There is a further advantage in being located on a good road, or near the electrics, convenient to a good community, which is, and is not, a business advantage, according to how you look at it. It might be called more properly a social advantage, but is it not good business sense for one to so locate himself that his children may have the benefit of first class schools without inconvenience? So that the whole family may have the opportunity to attend good churches, good entertainments, to participate in social affairs of a higher grade than those usually common in distant country sections? And will not the mistress of the household be better contented and take more interest in the farm, if she can, when she feels like it, drop into the town to visit, or to trade, or to otherwise mingle with other people?

The days when to be a farmer means to bury one's self in seclusion have gone by. A farmer is just as much, or should be just as much, a part of the community as anyone else in the community. Farming is an honorable profession, and when carried on with good judgment and discretion, is a profitable one. We believe that the poultry man who locates his family, so that it will have at least the ordinary advantages of modern civilization, will in that respect at least, be making an investment that will yield him excellent returns.

TECHNICAL CONSIDERATIONS

Of course, what we may term technical conditions should be carefully considered in the selection of any farm. The matters to be looked into are the nature of the soil, the slope of the land, the drainage, the elevation, exposure to winds, water supply, etc., etc.

RULES FOR SOIL

Perhaps the most important thing to investigate, is the nature of the soil. Poultry cannot be raised successfully on a heavy, clay soil without considerable expense; that is, the poultry will not flourish on such soil unless it is frequently turned over, cultivated and cropped, and unless it is so graded and drained that it will not contain too much moisture.

"Cleanliness and freedom from moisture must be secured if the greatest success is to be attained. Without doubt filth and moisture are the causes—either directly or indirectly—of the majority of poultry diseases, and form the stumbling block which brings discouragement and failure to many amateurs. It must not be inferred that poultry cannot be successfully reared and profitably kept on heavy soils, for abundant proof to the contrary is readily furnished by successful poultry men who have to contend with this kind of land. The necessity for cleanliness, however, is not disputed by those who have had extended experience in caring for fowls, particularly the less hardy breeds."

"When the fowls are confined in buildings and yards, that part of the yard nearest the buildings will become more or less filthy from the droppings and continual tramping to which it is subjected. A heavy or clayey soil not only retains all of the manure on the surface, but, by retarding percolation at times of frequent showers, aids materially in giving to the whole surface a complete coating of filth."

It is an advantage to have soil that is porous and well drained, a sandy or gravelly soil with an open sub soil offering most desirable conditions. Nothing, however, is gained by going to extremes in this matter. Because dryness is desired is no argument that it is good practice to build on a barren sandbank, as is sometimes done. Such soil is unnatural. Entire lack of moisture is harmful to fowls: their feet and legs become more or less shriveled, cases of lameness occur, and in summer the intense heat reflected by the bare sand causes much distress. An excellent rule for general application is to select a farm where the soil is well drained, but sufficiently moist to support a good sod.

Farms are particularly undesirable that consist of low land on the banks of streams or ponds, especially in narrow valleys where fogs are common and where the poultry houses must lie in the shadow until late in the morning.

AVOID EXPOSED LOCATIONS

A farm is also undesirable that is unusually exposed to storms and heavy winds, which are unpleasant to fowls and under which conditions it is most difficult to control the ventilation in the poultry houses, especially houses of the open front construction.

On the other hand, care should be taken to select a farm that has good air drainage. Many farms which are apparently well suited to poultry are deficient in this respect. They appear to be really pockets where damp, cold, stagnant air settles and in which fowls are uncomfortable, and most susceptible to disease. In such places frosts are noticed first, fog lies until late in the morning and in the summer the air is often oppressive and intensely hot. Such conditions are as bad for other lines of farm work, crops, orchards, etc. as they are for poultry.

Therefore, while protection from storms and heavy winds should be a strong consideration in selection of a farm, natural windbreaks being taken advantage of to the fullest extent, one must not go to the other extreme and select a location where there is not a reasonable circulation of air. It is a good plan to visit the proposed farm after nightfall, in order that it may be noted whether or not the damp air drains away. We believe that it will in most cases be wise to look for a location on a ridge or elevation that is, to quite a considerable extent, above the level of the surrounding country.

SOUTHERN SLOPE

To be perfectly adapted for poultry keeping the farm should have at least a gentle slope to the south, or to a point a little east of south, in order that the houses may have the advantage to the fullest extent of the winter sunshine. Besides furnishing warmth and cheerfulness to the fowl, sunshine also serves as an indispensable germicide. So it is not enough that the land should be located on a gently sloping ridge. The purchaser must take pains to see that that part of the ridge that slopes to the south or southeast is land that will be well adapted for his purpose.

Care should also be taken, of course, that the farm is amply supplied with good, pure water.

TO SUM UP

To sum up the matter as to the location and nature of the farm to buy, the ideal farm would seem to be one that has a soil of sandy loam, located on a ridge, slightly above the surrounding country, sloping gently to the southeast, and protected in the direction of the prevailing winds by higher land, woods land, or some other natural windbreak.

CLIMATE

Very few of us, who are considering entering the poultry business, would be in a position to change our location to such an extent that, if we were located in a climate to some degree unfavorable, we could move to some other distant climate that might have more advantages. It cannot be denied that climate has some influence upon the profitable production of poultry and eggs. But the influence is much less than many writers claim to be the case.

Cold winters in the north do not furnish ideal conditions for low cost production, neither do the long, hot summers in the south; but neither of these conditions are obstacles that it is impossible to overcome, nor do they affect profits to such an extent that it would be wise to make a radical change on account of them.

DON'T EXPECT PERFECTION

When the new poultry man begins to look for the place that will be suitable for his venture, he will likely find that no place will combine all of the advantages. He will have to consider carefully the advantages and disadvantages of all the farms he looks at, and weigh them all in the balance. His chief anxiety should be to see to it that no location is selected that is open to serious objections.

FARM CONDITIONS BEST

Many of our readers will, no doubt, have a strong inclination to make their initial venture in the back-yard of some city home. We cannot deny that a small flock of hens can be successfully kept in this location, and that the experience gained will be of much value.

The natural place for poultry, however, is the farm. The open range, cheaper food and low cost of labor and living make this the surest place to lay the cornerstone of a successful poultry business. Here fowls can be kept in large numbers at the lowest practicable expense, while the farmer can secure larger production, maintain as high a standard in quality and market his products to practically as good an advantage, as he can from any kind of a location within the city limits.

There have been comparisons of costs made by government experiment stations as between the expense of keeping farm flocks and town flocks, in which it was shown that the average farm cost was 89 cents per hen, and the average town cost $1.57. While these figures would probably fail to reflect accurately the cost today on account of the generally higher level of all commodities, we see no reason why the ratio should not remain the same.

If there is one recommendation that we would make beyond all others to the beginner in the poultry business, it would be to make his beginning on a farm. Should his first venture made on a city lot result in poor success, it is extremely doubtful if he would try again elsewhere; whereas if the same energy and thought had been expended under farm conditions his chances of success would have probably been good.

AN IDEAL RANGE SHOWING 4 X 8 COLONY HOUSE

CHAPTER THREE

CHOOSING THE BREED

It is neither feasible nor desirable that we go into a complete detailed history of each breed in this book. Our purpose is to give practical information to the beginner. It is of very little importance to him how any breed originated, or how it was developed during the hundred years previous to his entering the business. Later on, when he becomes sufficiently interested to desire more complete information regarding these matters, he can find it in a number of valuable poultry books that cover the subject.

We assume that the matter of most interest at this moment is as to whether this breed or that breed lays a light or dark egg, a large or small egg, many or few eggs, whether the size of the bird is large or small, whether the hens are good foragers, large or small eaters, etc., etc. These matters we shall take up in as comprehensive a way as is consistent with reasonable brevity.

The selection of a breed is not of quite so much importance as it seems to the beginner. Of course we thoroughly believe in any poultryman keeping that breed that most strikes his fancy, because he will be more interested in his work under that condition. It is probably true, however, that there is no one breed that is noticeably superior to any one of a dozen other breeds from a practical utility standpoint. It may be true that there are a dozen breeds that will give better average results than some other dozen, and it is also undoubtedly true that many breeds are of value only from a fancy standpoint. Some breeds are only of value to satisfy a curiosity in experimenting. Breeds of these latter types of course should be avoided by the poultryman whose object is to make money.

STRAIN IMPORTANT

Personally, we feel that the opportunity for fortunate selection of desirable birds lies more in careful consideration of the strains of the breed, than in consideration of the different breeds. If the beginner will allow his fancy to dictate what breed he shall keep, and his judgment to dictate the strain, he will very likely come out alright. The difference in strains in any breed is entirely due to the degree of care used in developing the strain in selection of breeders, maintainance of health, size and laying qualities. These are the factors that influence profits.

Environment will, of course, play an important part in the selection of the breed. One will naturally desire to select the breed that will fit in best with his conditions. By conditions, we mean the size and nature of the farm; to what other business the farm is devoted; what the local market will demand in products; whether a brown egg or white egg is required, or yellow or white poultry, etc.

ONE BREED, BUT THE RIGHT BREED

This much is certain, however, if the poultryman intends to make a specialty of eggs and poultry meat, or of eggs alone, he should select a good utility strain of a good utility breed. If he desires eggs alone, it will no doubt be good judgment to select one of the lighter, more active breeds, represented by Mediterranean or Hamburg classes, (members of classes enumerated below) that are heavy egg producers, "non sitters" and that are less expensive to keep, rather than one of the heavier breeds that sacrifice their egg producing ability somewhat to the ability to produce meat. If, on the other hand, his main object is to raise choice roasters or capons, he will do better to select one of the very heavy breeds, (Asiatic class, or heavier varieties in American class) which are, however, too inactive to be very prolific layers. And then again, there are those breeds, classified as "American Breeds" that combine both qualities to a very reasonably satisfactory degree.

It will be found to be good practice to pin one's faith to one breed, to two at the most, or at least to one or two at a time. Each breed requires different care in many ways.

The confusion incident to having two or three different systems of care for two or three different kinds of birds usually leads to entire lack of system in the whole operation with consequent unsatisfactory results.

Fowls which are similar in type and breeding are most easily handled on account of their greater uniformity and character.

NO REAL PROFIT IN CROSSES

It is an axiom among those who have had experience in the business never to cross two breeds or to spend any time on mongrels. Sometimes a cross proves successful for the first year, but the offspring from the cross cannot breed with anything but unsatisfactory results. To have as a permanent flock any particular cross that has proven profitable means to revert in breeding each year to the original thoroughbred stock. As the thoroughbreds may be said to be always as good layers as any cross that can be produced by them, we can see no object in going through the trouble and confusion of producing this cross each year. Furthermore, when one wishes to dispose of any of the cross bred birds, he can do so only in the market as poultry, for they are of no value as breeding birds to anyone else.

Standard bred fowls, bred with a definite object in view, in practically all the popular breeds, lay more eggs than those of mixed breeding, or of no breeding at all. Their eggs are larger in size, more uniform in color, and the young stock grows faster and larger and makes a better appearance in the market. Furthermore the possession of a flock of fowl all of the same breeding, size and color should be a source of genuine personal satisfaction to anyone, and this will result in more attention being given to the flock, which in turn always means better returns.

BEGINNING WITH SHOW BIRDS

There is occasionally a beginner who has no interest in the utility end of poultry keeping; his chief desire is to become an exhibitor of fancy stock, to have the pleasure of showing birds against his neighbors in the many poultry shows that are held in the fall and winter of each year. His interest is entirely an aesthetic one, or a sporting one. It is particularly important that such a beginner should choose the breed for which he has the greatest liking, but should it happen that he has no special fancy for any one breed, his choice should be one of those breeds which are shown in greatest numbers in the territory in which he exhibits. He will find more pleasure in exhibiting in large classes than in small, and if his interest should extend further to the end that he wished to sell birds occasionally, he would find the demand greater for birds in these popular classes, than in the smaller classes.

It may not be out of place, however, to suggest that the beginner who starts in a fancy show bird business without any previous experience with poultry of any sort, has a hard and discouraging few years ahead of him. He may get his start in stock of this class by buying eggs from exhibition birds, or by day old chicks, or by buying a pen of the grown birds themselves for breeders. As the chance, however, for getting any reasonable number of really choice foundation birds from a reasonable number of eggs is very uncertain, and as very few breeders of exhibition stock sell day old chicks, it would seem to be best, in our judgment, to depend upon the purchase of breeding birds for a start. Great care should be exercised in selecting these birds. The beginner should get the advice of some man of experience whom he can trust; should know the pedigree and history of each bird; see the parents, and other birds of the same breeding if possible; because, unless the mating is made up in view of results that have already been obtained with stock of a similar character, the chances for good results from the mating are no more than even.

HEREDITY'S PART

Haphazard methods in breeding for any object, result in nothing. More especially is this true in breeding for exhibition purposes. The breeder, in seeking to secure transmission of certain characters from one generation to another, is dependent upon heredity. Oliver Wendell Holmes illustrates the nature and extent of heredity by saying:

"We are all of us omnibusses in which our ancestors ride."

The meaning of this in poultry breeding is, that each fowl in its character is the sum of its ancestors, having, in greater or less degree, traits inherited from each one back along the line to the original source of all birds. Qualities that have remained hidden through many generations may suddenly reappear with remarkable distinctness and vigor.

The statement that "like begets like" is subject to many seeming contradictions. Which of the characters of the parents will be active in the offspring depends upon a number of factors. It does not follow because certain characters are active in the parents, that they will also be active in the offspring. The sure way to make the characters desired the distinguishing character of the strain, is to make these desired characters so strong that they overpower all other characters.

Scientists have discovered many important laws governing breeding. In the light of these discoveries breeding is becoming more and more an exact science, and less a matter of blind chance. The breeder will be more successful in bringing out particular distinguishing characters if the fowls mated are of similar blood lines: in other words if they are related. This means inbreeding. And in inbreeding great care must be taken that the weaknesses of the fowls are not intensified in the offspring.

It is not our intention to go into scientific details regarding the breeding of exhibition specimens. We have gone perhaps farther than we should in this chapter, but only for the purpose of impressing upon the novice the difficulties he is likely to encounter, if he starts in without experience to build up an exhibition line of birds.

NINETY-NINE IN ONE HUNDRED UTILITY BREEDERS

As ninety-nine out of a hundred, however, of poultry beginners have in mind only the raising of stock and eggs from a utility standpoint, we shall devote ourselves almost exclusively to a description of the breeds desirable from this standpoint. There is one matter that in the past has rightly had a prime influence in the selection of breeds, and that is the color of eggs it produces. From time immemorial New England has been stubbornly partial to the brown egg; likewise the New York market has been partial to the white egg, and anyone keeping layers in either of these localities has had to keep these facts before him; otherwise he would sacrifice several cents per dozen on the amount he would receive for his eggs. During late years, however, the public has been gradually educated to the fact that the color of the shell is of no real significance from the standpoint of the food value of the egg, so there is perhaps less necessity of being limited by this consideration in the selection of a breed.

Of course every beginner has to get his new stock from some breeder already established. Even though we are breeders ourselves and may seem prejudiced to the casual reader, we feel that we are justified in impressing on the beginner for his own sake the necessity of buying good thoroughbred stock from a breeder of reputation. The beginner, however, must not allow himself to be too much influenced in his selection of breed by the statements of any one breeder, who will of course be prejudiced in favor of his specialty. We are personally prejudiced in favor of Barred Rocks, and Single Comb White Leghorns, each for a different purpose, but we do not wish to urge this prejudice on to anyone else. We keep other breeds besides the two that we like best, because we are in business to sell breeding stock and intend to keep all those breeds of merit that have any great popular following. You will note in the description given below that the first five breeds we have described are the ones that we breed ourselves. But we describe these five first for the same reason that we keep them, that is, because we have found them to be in most popular demand, and not because we have any desire to urge them upon the reader.

With the above preface to this chapter, we give below the principal characteristics of each of the most popular breeds. We will leave it to the reader who has sought this book for advice to select the breed that best meets his ideal. There are many breeds that we have omitted, because many of them are in no sense popular, some of them not even generally known about, and some with absolutely no merit.

DESCRIPTION OF BREEDS
GENERAL CLASSIFICATION OF UTILITY BREEDS

American Class—Plymouth Rocks, Wyandottes, Javas, Dominiques, Rhode Island Reds.
Asiatic Class—Brahmas, Cochins Langshans.
Mediterranean Class—Leghorns, Minorcas, Andalusians, Anconas, Spanish.
Polish Class—
English Class—Orpingtons.
Hamburg Class—

PRACTICAL CLASSIFICATION

For Eggs Principally—Mediterranean Class.
For Meat Principally—Asiatic and American Classes.
For Both Eggs and Meat—American Class.

BARRED PLYMOUTH ROCKS

If we take into consideration all sections of the country the Barred Plymouth Rock is probably the most popular variety of fowl, especially in the small flocks on American farms. The reason of this popularity, while partly due to the enthusiasm with which it was adopted by fanciers with consequent large classes in all the shows, is really, because of its excellent quality as a utility fowl, having both good egg producing ability and good medium size for meat purposes. The chickens grow fast and come to laying in a very reasonable period. In isolated instances it has been known to lay its first egg in ninety days; many specimens mature to laying in four months, while the flock as a whole will usually be well under way in egg production in from six to seven months. This quality is valuable because it avoids the necessity of hatching too early in the spring to be sure of fall eggs when prices are high. The instances of extremely early maturity are of no special value, except as indicating the tendency of the breed or strain to mature within a reasonable period. As a matter of fact, it would be suicide to use these extremely early maturing birds for breeding because they usually lack in size.

VERY HARDY

The Barred Plymouth Rock is a very hardy bird, especially well adapted to the cold, northern climate. They are well feathered, have low combs and are compactly built. They stand confinement well, but when allowed their freedom, prove excellent foragers. The color of the egg layed by this breed is a moderately dark brown, and the eggs are of good shape and size. Under conditions that are at all favorable, they are excellent layers, in fact a great many of the best egg records have been made by the hens of this variety. The average price received for Barred Plymouth Rock eggs in the course of a year will run as high as that received for the eggs of any breed, on account of their early maturity which brings good production during the period of high prices.

Plymouth Rocks, of whatever color, in their best development are long in the back and have a good long, even keel. The breast is broad deep and well rounded: the abdomen strong and full. The head is large, rather round, prominent in front with broad crown, and carried well up. The eyes are bay in color, large and clear. The comb is well set, slightly below medium in size and has five serrations. The beak is short and stout, yellow in color. The legs are yellow, as is also the body, making it especially desirable for meat purposes in the market. But of course, it is the broad, deep and well rounded breast that gives these fowl their value as table poultry. The color of the Barred Rock is a bluish gray, barred with narrow parallel lines of dark blue approaching almost to a positive black. In thoroughbred specimens the barring must show the entire length of the feather and be close in all sections except in the primary and secondary feathers. The standard weight of the Plymouth Rock is as follows:

```
Cocks       Nine Pounds
Cockerels   Eight Pounds
Hens        Seven and One-Half Pounds
Pullets     Six Pounds
```

These weights apply to all varieties of Plymouth Rocks.

WHITE PLYMOUTH ROCKS

White Rocks have nearly all the characteristics of Barred Rocks. They are possibly somewhat less hardy, less easy to raise, and the eggs do not hatch as well as Barred Rocks within five per cent. Although the Standard of weight is the same as the Barred Rocks, they run a little heavier at all ages, doubtless due to the special selection of large birds for breeding by the early breeders of the variety. Personally, we believe that White Rocks look better in the larger size than the Barred Rocks. For meat purposes the White Rocks are often preferred to any other breed; and justly so, because they are of excellent size, have the standard Plymouth Rock breast and look especially clean and attractive when carefully dressed. The pin feathers do not show because of their color.

There are several breeders on the South Shore in Massachusetts who have made a specialty of raising and marketing what are known as soft roasters. In the early stages of their business we believe that Brahmas were extensively used, but, as the business developed, experiments were made with other breeds with the result that both Barred and White Plymouth Rocks became popular for the purpose, more especially the latter, on account of their color. The White Rock, we understand, is now used very nearly exclusively by the best raisers of this kind of poultry. The reason for the substitution of the

White Rock for the Brahma, or larger variety, seems to have been that, while, of course, the Brahma would eventually grow to a larger size, the White Rock would reach the size where it was marketable more quickly, so there was consequently more profit.

The egg layed by the White Rock perhaps never runs quite as dark brown as some specimens of the Barred Rock eggs, but we believe they will run more uniform in color, and that they average somewhat larger in size, no doubt due to the larger average size of the birds. Both Barred and White Rocks make good broilers and reach the broiler size usually at about ten weeks, but neither variety is perhaps quite as popular for this special purpose as some of the other breeds.

It may be interesting to the novice to know that the White Plymouth Rock was originally an offspring from the Barred variety,—what is known as a sport. Many more or less indifferently bred strains of the Barred Plymouth Rocks even now throw these sports occasionally. The reason of it is, no doubt, that the Black Java was part of the original foundation of Plymouth Rock stock and all solid black fowls are apt occasionally to throw white chicks. Then again, the cross of the White Asiatic fowl, made for the purpose of enlarging the size of the frame and the egg both, and for clearing the plumage, might be expected to crop out in the offspring. We must not have the impression that the first of these white sports had good clear color: it took several years of great care in mating to get them started toward breeding true.

SINGLE COMB WHITE LEGHORNS

Although there are many varieties of leghorns, the popularity of the Single Comb White Leghorn is so very great that one thinks only of them when Leghorns are mentioned. At one time the Brown Leghorn was considerable of a favorite, but its reputation, especially as a utility fowl, has been very much overshadowed by its sister, the White Leghorn. And there is good reason for this, because whereas the Brown Leghorn is not any better in either size or laying qualities than fifty years ago, the White Leghorn has made great advancement in both these respects. The standard weight of a Leghorn pullet is three pounds, but the poultrymen nowadays, who use pullets for breeders that weigh less than three and a half pounds, are the exception, rather than the rule.

Even with its increased weight, however, the Leghorn in no sense can be called a meat fowl, that is, a fowl that it is profitable to raise as a roaster or capon, because the maximum weight of this species is less than is required for a marketable roaster. Many plants, however, make no point of raising poultry for meat purposes; they confine themselves to the production of eggs, disposing of surplus cockerels and old birds for whatever they can get for them as live poultry. The young cockerels are sold as broilers. We believe it is fair to say that Leghorns make as good, if not better, broilers than any other breed. This is due to the rapidity with which they grow to a certain size, and to the type of the bird. Pretty much all of the surplus flesh that a Leghorn puts on is found in the breast. The shape of the bird adapts itself particularly well to the requirements of a good broiler.

For this reason, even though the breed is not a meat breed, strictly speaking, it still offers the opportunity to be marketed in this way, and there is no lessening in the profits received as compared with cases where other breeds are used, provided the cockerels are not kept beyond the broiler stage. There is a loss, however, in disposal of old birds, as the weight runs hardly more than fifty per cent of most of the American birds. This loss however, is more than offset by other circumstances connected with the keeping of Leghorns'

The White Leghorn is a particularly desirable breed for the poultryman to select, whose main object is to raise eggs for market. There have been times when this would not be so true in the New England states as it is today, but as we have said elsewhere, the New England public is beginning to realize that the color of the shell of the egg has no real significance. A large percentage of the big commercial plants in the United States are stocked with White Leghorns, and we believe the results from them for this purpose have been very generally satisfactory.

One of the reasons for this is, no doubt, the less expense of maintaining them. Leghorns are a nervous, active fowl, continually foraging, so that the food that they consume is easily assimilated and transposed into eggs. In other words, their propensity for exercising keeps them in first class physical condition, so that they easily digest the food given to them. This same activity prevents them from becoming too fat, which is one of the most serious hindrances to good egg production. Then as a final consideration, it is claimed that, owing to its size, the Leghorn consumes very much less food than any of the other popular egg producing breeds. In fact, it is often stated that one hundred and fifty Leghorns can be fed for the same cost as one hundred of the medium sized breeds. The food given a mature hen is utilized in three ways in approximately equal amounts, viz:

>One-third for maintaining the body
>One-third for producing eggs
>One-third is wasted directly or indirectly

It would seem logical that the lighter breeds would require less food to maintain the body, would consequently make less indirect waste and would require no more food to produce a given number of eggs. Of course lighter breeds are more active but this activity induces better assimilation of food given them.

It is also claimed that one hundred and fifty Leghorns can be housed in the same space as one hundred of the other breeds, so that the Leghorn offers a saving in investment as well as in operation. We are inclined to take some exceptions to the statements regarding the comparative number that may be housed in a given space, but it is without question true, that a certain number more may be housed than of the larger breeds, probably twenty to twenty-five per cent more.

It is these economies in the keeping of Leghorns that offset the difference in meat value. Whether to the full extent or not, depends much on circumstances, but we are willing to coincide with the general opinion, that, as a purely egg market proposition, the Leghorn is probably the most desirable of all breeds. They are, perhaps, not as desirable in the colder, northern states as in the more temperate states, but the difference is not so great as it used to be, now that a different style of housing is in vogue. Our own experience has shown that the Leghorn stands the northern winters in the types of houses that we use fully as well as any other breed. In general practice, also, Leghorns may be kept in larger flocks than larger breeds, so that simpler, and less laborious methods may be used in their management. This allows a still further saving, and it is all these small savings together which makes the Leghorn so desirable from the egg plant standpoint.

Because of the fact that Leghorns are great foragers, it is generally believed that they should be kept only on farms where there is large available area. We do not think that this is strictly true. The experience on our plant, where we house the birds in the twentieth century type of house, is that the Leghorns will stray no farther, if as far, away from the house as the Plymouth Rock or Rhode Island Red. Leghorns do object to being confined in too small a space, that is, when confined in such space they will fly over any ordinary obstruction to get out of it. At the same time they will stand confinement with good health as well as any breed, and may be profitably kept in small yards on city lots, provided they are confined with fences sufficiently high.

Another advantage of Leghorns for the beginner to consider is the fact that the eggs hatch, as a rule, several per cent better than those of any other breed, so that the young chicks may be obtained each spring either by one's own raising, or by purchase, cheaper than those of any other breed. This makes, of course, some considerable difference in the investment and renewal cost.

As to the general shape of the Leghorn fowl, the male should be trim, active and graceful: its body plump and round, broad at the shoulders, and tapering toward the tail. He should have a hard, close fitting plumage and the tail should be well balanced on a fair length of shank and side. The breast should be full, well curved, prominent and carried well forward; neck well arched and head carried erect; back of medium length, with saddle rising in a rather sharp concave sweep to the tail. The legs are bright yellow, the beak yellow, the eyes bright red, the comb straight and firm.

The Leghorn hen is not as graceful as the cock but may appear more sprightly.

Leghorns are classed with the yellow meated fowl, so in this respect are good poultry. The egg of the Leghorn is chalk white in color, which may be an objection to the keeping of the breed in some special localities. In a good strain of Leghorns the eggs are very excellent in size, and where they are packed separate from eggs of other colors, they appear attractive and appetizing. There is a tendency sometimes for the eggs to have a slight tinge of pink, but the true breeder will attempt to breed this out, so that his product may be uniform in color as well as in size.

SINGLE COMB AND ROSE COMB RHODE ISLAND REDS

The Rhode Island Red one is of the most practical breeds from a utility standpoint, and one of the most desirable from the fancy standpoint. Its particular combination of color is exceedingly difficult to breed with any great degree of perfection, but for this reason is the more interesting.

Few new varieties of fowls have enjoyed a more quickly gained prominence than the present up-to-date Rhode Island Red. The popularity is about equally divided between the Rose Comb and the Single Comb varieties. The classes of each in all shows are of about equal size, and there are probably as many utility plants stocked with one as with the other. In this brief description, we shall not differentiate between the two, as we are dealing primarily with utility qualities and there is no difference between the two in this respect.

Rhode Island Reds are primarily utility birds and a general purpose fowl. They were brought into line by breeders known as utility men. Breeders of the variety claim that they are better layers than the other American varieties, that they are the equal of the leghorn in this respect, and they claim, further, that the carcass has a small percentage of entrails. But one must take with more than one grain of salt the claims of the ardent admirer of any breed.

In reality the Rhode Island Red is similar to the Plymouth Rock in its general characteristics. The standard weight of the breed is about one half pound less than of the Plymouth Rock and in general practice it is bred fully as much lighter as this; considerably lighter in fact than the White Rock. The vigor of the Red, however, is on a par with that of any breed. The eggs hatch reasonably well and the chicks are not hard to raise.

The laying qualities are excellent. To our mind the Red excels almost any other breed in fall and winter laying when prices are high. It matures rapidly, rather earlier than other breeds, and from its first laying until the early spring, it is a persistent producer. The chief trouble with the Red as a utility bird appears after the early spring arrives, when it develops a very persistent desire to be broody, and it is less easily broken up from its brooding than other breeds. The females make splendid mothers, however, on this very account, but this is not a matter that interests large commercial plants to any great degree. On the contrary, the tendency to brooding is a distinct disadvantage, because it curtails the egg production, and creates considerable seemingly, unnecessary labor.

We think it will be conceded that the Red lays a more handsome egg than any other breed. It is a very beautiful, rich, dark brown, and as a rule runs unusually good in size. If any eggs were to demand a premium in price, we should feel that the Reds deserved this advantage.

The Reds also make nice poultry. They are another of the yellow skin variety, have good plump breasts and a good distribution of meat in other sections. They mature quickly as broilers and are in good demand for this purpose. Like other American breeds they fatten easily if overfed, but if good judgment is used in feeding, this is a difficulty that may be easily avoided.

The standard weight of the Red is:

Cock	8½ pounds
Cockerel	7½ pounds
Hen	6½ pounds
Pullet	5 pounds

The general shape of the breed is rather shorter and more compact than of the Plymouth Rock. One would judge from this that the egg producing capacity was not as good as that of the Plymouth Rock, but as for this we will not venture to say. The back is broad and in the main nearly horizontal, the breast is broad, deep and carried nearly in a line perpendicular to the base of the beak. The body is deep and broad, keel bone straight and fairly long, extending well forward and back, giving the body an oblong look. The thighs are large and of medium length. The comb is medium in size, the beak short and regularly curved, the head of medium size and breadth. In color the general surface should be a rich, brilliant red, except where black is desired. The birds should be so brilliant in lustre as to have a glossed appearance. Black is desired in the underweb of the wing flights, in the main tail feathers, greater tail coverts, and in the two main sickle feathers. The female is rather lighter in color than the male, and is not as brilliant in lustre as the male.

One of the most unsatisfactory features in a flock of utility Reds is the lack of uniformity in color. The beginner should not be too much discouraged if he finds this trouble in his flock, because he would be most likely to find it in any flock that he would take the time to inspect. It is no doubt due to the fact that Reds are of such recent origin, no one shade of color has as yet become firmly fixed in the breed, but very great improvement has been made in this respect during late years, for which the ordinary Red breeder has the fancy breeder to thank.

WHITE WYANDOTTES

When we consider that in 1883 there were absolutely no Wyandottes, we wonder at the extreme popularity of this breed today. There are very few shows in the country where the Wyandottes are not among the two or three biggest classes exhibited. This is more particularly true of the White Wyandotte, whose popularity very much overshadows the popularity of the other varieties of the same breed. There are several other varieties, more prominent among which are the Buff Wyandottes and Columbian Wyandottes, neither of which, however, need to be considered seriously in the class of popular breeds from a utility standpoint. But for the purpose of this chapter, it is not necessary for us to argue this point, as the practical description that we shall give would apply equally as well to those varieties as to the White.

In size, the Wyandotte is supposed to be practically the same as the Rhode Island Red, but in actual practice we believe it will be found to be somewhat smaller, due of course, to the methods of breeding that have been applied to the breed in the past. The standard weight calls for the following

Cock	8½	pounds
Cockerel	7½	pounds
Hen	6½	pounds
Pullet	5½	pounds

Except as this somewhat lighter weight causes a loss of income when the old hens are sold off, it is not a matter of especially great importance, because the Wyandotte will readily reach the size desired for roasters, and do so quickly. For broiler purposes the Wyandottes have no superior. It is termed a "bird of curves," which means a body well rounded in all sections. This is particularly true of the breast, which is nearly as plump and full as that of the Leghorn. The Wyandotte is meaty at every point. The color of the meat and skin is of that attractive yellow so much in demand in our markets. The breed has the combination of attractive shade and color, and the white plumage removes all chance of prejudice so often advanced against dark pin feathers, thus giving it three very important advantages for sale as market poultry. It is in some respects the model for market poultry.

Wyandotte breeders claim much for their variety as egg producers. It would be a question in our minds if Wyandottes would average as well as some other breeds in this respect, and we say this in face of the fact that the best egg production record we had this last season among our flocks was from one pen of our Wyandottes. We base our opinion more especially upon the general shape of the bird. It does not seem to us that it has the capacity for production that some of the other breeds possess. Many Wyandotte flocks that have made excellent records in egg production have been found, upon investigation, to be more nearly of Plymouth Rock type than of Wyandotte. As we have said, however, elsewhere in this book, we believe strain is more important than breed, and there can be no doubt that there are several strains of Wyandottes which are very excellent egg producers.

The color of the Wyandotte egg is similar to that of the Plymouth Rock, running perhaps somewhat lighter in color, but averaging well in size. Taking our experience as a basis, Wyandottes have some disadvantages as producers. The chicks are not as hardy as those of some other breeds and are more difficult to raise. We cannot, however, criticise the vigor and health of the birds when grown.

ORPINGTONS

The Orpington fowl originated in England and is without question the most popular breed in that country. There are several varieties, including the Black, the White, the Buff, the Jubilee and the Spangled. It was first imported into this country about sixteen years ago, and has increased rapidly in general popularity since that time.

The Orpington was fortunate in being vouched for here by men of more than the usual enterprise. The breed undoubtedly has excellent utility merit, but we doubt if it is in such a degree that it would have gained such popularity in such short time, if it had not been for the excellent advertising that was given it by the original importers. Today the Orpingtons comprise a very material part of the entries of all shows. The Black, the White, and the Buff Orpingtons are undoubtedly the most popular varieties. The birds are large sized, the standard weight of the breed being as follows:

Cock	10	pounds
Cockerel	8½	pounds
Hen	8	pounds
Pullet	7	pounds

When the Orpington was first produced, it was a very poor layer, owing to its development by injudicious inbreeding. Through the efforts, however, mainly of one man, its deficiencies in this respect have been very much overcome, and it may now be considered satisfactory in this respect. The quality of the egg layed is particularly fine, the color being an excellent shade of brown and the size unusually large.

On account of its size, the Orpington makes a desirable meat bird, and it grows reasonably quick to the marketable ages. The quality of the flesh is also good, but in some markets it has the drawback of having both white legs and white skin, which we cannot bring ourselves to believe is nearly as attractive for poultry as the rich yellow of some other breeds.

Much is claimed for the vitality of the breed. Judging from its ancestors, it is surely entitled to have good vitality, but so far as we know, the hatchability of the eggs and the livability of the chicks are no better or no worse than in the case of the average breeds.

In shape the Orpington may be described as a blocky bird. Its back is shorter in proportion than that of the Plymouth Rock, but it stands high and has a very nicely rounded breast. We should consider that its type was better adapted to poultry purposes than the Plymouth Rock, but not as well adapted to egg production. In general, however, we can fairly say that the type is satisfactory for a general purpose fowl.

BRAHMAS

The Brahma is one of the oldest and best known of our breeds. While it is by no means as popular today as it was in the preceding generation, it still has very many ardent admirers. Its loss of popularity with the general farmer was entirely due to the fact that other breeds appeared that were better general purpose fowls and that yielded more profit.

From a strictly market poultry standpoint, speaking from the standpoint of its size and quality at maturity, no breed is superior to the Brahma. It has been replaced on market poultry farms by somewhat lighter breeds, only because these lighter breeds will reach the size required in somewhat shorter time and at less expense, and also because, when these lighter breeds do reach this size, they are well filled and rounded out, whereas the Brahma is somewhat short of this condition. The truth of the matter is, that the standard weight of the Brahmas is unnecessarily great. Of course, when a matured Brahma is sold for poultry it brings more money than the other breeds, but the cost per pound is higher, so that the net profit is probably no more. We ourselves, are very great admirers of the Brahma on account of its fine size and vigor, but it is not popular enough for us to keep as a sales proposition, and were we looking for a good, utility breed to keep on a general farm, we would undoubtedly sacrifice our liking for this breed to the greater utility merit of some other breed.

There have been many nice egg records made by the Brahma. It is not generally considered, however, to be one of the good egg producing breeds. It is too large and inactive for this purpose. The quality of the egg layed, both in color and size, is probably superior to that of any other breed, but the numbers of them are too few. If eggs were sold by the pound instead of by the dozen, the Brahma would be in a better position.

There are two common varieties of Brahmas, the Light and the Dark. Our preference is for the Light variety, with its pure white body feathers set off by black striping in the hackle and tail; it is very attractive. The shape of the Brahma is about as near a block as one could get it. It seems to us to bear the same relation to other fowl that the truck horse does to other horses. Its back is broad, its breast very prominent, deep and full, well rounded out, and it has full, prominent abdomen and sides. It is a genuine meat bird and if it would reach its maturity more quickly and at the right size, it would be ideal for this purpose.

MINORCAS

The Minorca is another breed that is more or less in favor. It is somewhat the same type as the Leghorn, and we believe that if it had been bred with the same skill that it would be quite as popular for utility purposes as the Leghorn. It lays an egg very much like the White Leghorn, it being of very nice size, and good, white color. We have known flocks, also, to make unusually good records in the yearly productions, but the variety is not commonly enough raised to have a very great reputation as a good choice for market egg farms. Nor under the conditions that exist would we venture to recommend the breed for a commercial egg plant. The average ability as a layer is not sufficiently well established by several generations of careful breeding, and to our mind the white flesh and skin of the bird is against it as a poultry proposition. Again, the combs and wattles are too extremely large for northern climates.

The Minorca is a somewhat larger bird than the Leghorn but it has the same type of breast, and its meat is fine in texture and short in fiber, so that it is excellent for broilers, except in those localities where there is objection to the white skin and legs.

WHITEFACED BLACK SPANISH

There were formerly two varieties of the Spanish fowl, the one known as the Whitefaced Black Spanish, the other pure White with a white face. The White variety was never popular, chiefly because it was not considered so attractive. Judged from the utility standpoint the Black Spanish is more or less of a Leghorn type of fowl. In its history it has been both a popular exhibition breed and a popular general purpose fowl. At one time it was very commonly kept on the small American farms, but of late years it has stepped down and out of the list of general purpose fowls and is now considered almost entirely as an exhibition breed.

We believe if the Black Spanish should receive the same attention in breeding for utility purposes that has been accorded the Leghorn it could be made to be, within a reasonable number of generations, one of our best egg producers. It has the advantage of being somewhat larger in size than the Leghorn, but it has a decided disadvantage from a poultry standpoint of having blue legs and bluish-white skin. Its type is very similar to the Leghorn, having an excellently rounded breast, so that it makes a good broiler. The color of the eggs is white.

ANCONAS AND HAMBURGS

These two breeds may be considered as further good examples of the non-setting egg producing varieties, like the Leghorns. They have the same deficiency in weight as the Leghorns, but both can likewise be successfully raised for broilers. One could hardly consider either of them in the same class as the Leghorn for commercial egg plant purposes, but we believe that this is due largely to the lack of careful breeding with this end in view in the past. The color of the egg is white and the eggs are not particularly large, but this matter of size is also due to a lack of effort in breeding.

In the Hamburgs again we have the same objection of blue legs and blue skin, or at least to us it is an objection. We presume that there are markets where this color is preferred in dressed poultry. But to our mind it is not nearly as attractive as the yellow legs and yellow skin. The Ancona is not subject to this same criticism, because the standard calls for the yellow color in shanks, feet and body, but it is not in reality as nice, bright yellow as the Leghorns are favored with.

Both the Hamburg and Ancona have the advantage of being strong, vigorous breeds whose eggs hatch well and whose chicks live well. While neither of them can be said to be an ideal utility bird today, we believe that they could be brought up to the proper standard should anyone take sufficient interest to attempt it.

There are other breeds which have merit in one way or another, like the Andalusians, the Polish, the Cochin, but as none of these would be selected for a utility proposition, we will not devote any space to them in this book.

CHAPTER FOUR
POULTRY HOUSES

Neither the cost nor design of a good poultry house need be a serious matter if one understands what the fowls need and for what purpose the house is being built. It may be said that there are three essentials in poultry house construction:

1.—The protection and comfort of the fowls
2.—Simplicity and economy in design
3.—Convenience for the attendant

The house that fulfills these three requirements is likely to be a satisfactory structure.

PROTECTION AND COMFORT

It is absolutely necessary in successful poultry keeping that suitable houses be provided for the flocks. Fowls cannot be expected to give results if they are left exposed to cold and storms in open sheds and in insufficient barns. At the same time an unsanitary house is worse than none. Comfort to the hen means something entirely different from comfort to the hen's owner and it is important to discriminate in the matter. For example, it would no doubt be more comfortable for the poultry man working in a house if the house were nicely warmed. But warm houses are the abomination of poultry keeping as far as the hens are concerned.

Warmth is a relative term, and while it is easily possible to go to extremes in constructing cold houses, fowls will be quite comfortable at very low temperatures if other conditions are right. At one time, in order to get winter eggs, it was thought absolutely necessary to have the hen house tightly shut in to retain the warmth from the hens: in many cases direct heat was supplied by stoves or steam plants. To-day we have better production and better birds with houses that not only have no heat but that are open the year round to receive the fresh air and sunshine. The necessity for providing summer conditions in the winter is an exploded theory no longer seriously considered with reference to housing poultry.

COLD HOUSES BEST

Cold houses, with open fronts, are best for adult fowl, and we have found them much better to a certain degree, for our young chicks. It is possible, however, to overdo the matter where winters are extremely cold. Houses in which combs are badly frosted are too cold for practical use. The house should be so arranged that it may be partially closed in extremely cold weather and made quite open in warm weather, but good ventilation free from drafts must be possible at all times. The temperature of good houses may be kept within reasonable limits in bad weather by restricting ventilation.

DAMPNESS DISASTROUS

Particular care must be taken that the birds are protected from dampness. This matter is very easily overlooked, especially by the novice, as it is acquired in so many ways. It is not sufficient that the house should be built on high, well drained ground, although it is important that these particulars should be looked after. Very often high ground is not especially dry, and unless it is well drained under the concrete or dirt floor, dampness will come up through. Either a dirt or concrete floor should be built up one or two feet above surrounding level. If a board floor is used, it should be placed equally high above, and the dirt beneath the floor thoroughly drained as well. In the average house we believe the board floor is preferable to the dirt or concrete, unless the conditions are ideal. In open front houses, in climates where the ground freezes solidly, the ground inside the open front house will not freeze as much, and in the warm days of the winter will thaw, making a certain degree of moisture that permeates up through the floor into the litter, making it damp, which is very injurious to the birds. About the only safe way to avoid this, is to build an underpinning about the whole pen to a point below the frost, so that the ground under the pen will not freeze at all.

Dampness in the litter is also acquired in shed roof pens by frost collecting during the night on the inside of the roof and dripping down into the litter during the heat of the day. The only way to prevent this is to build a regular board ceiling in the pen, or better still, to build a straw roof as we use in our Twentieth Century houses. This latter method is preferable, because it furnishes excellent ventilation.

Dampness is caused also by overcrowding a pen with too many birds. The moisture from the breath of the birds and from their droppings will soon make an overcrowded pen almost unlivable. At any rate, it makes it so unsanitary, that no bird will do well under those conditions. The litter in a pen will also become damp by careless handling of the water in a pen, or by carelessness in adjusting curtains on days when the rain or snow is beating into the pen. It pays to be forehanded and careful about such matters, because the birds will not do well on damp litter and it is cheaper to look after these little things than to change the litter unnecessarily. A damp poultry house will be a failure, no matter how carefully it is planned or how much expense is put into it. Fowls in damp houses are uncomfortable, unproductive and certain to become diseased.

THE THIEF IN THE NIGHT

It should not need saying that fowls must be protected against natural enemies like hawks, skunks, weasels, foxes, etc., but many poultry men are unthinkably careless about this very thing. Many a farmer will stand by and see his chickens carried off by crows and hawks, when the setting up of lines of ordinary string, the hanging up of a crow, or even the judicious use of a shot gun, will save him all this loss. Protection from the four footed animals, that steal into the hen house at night, is readily secured by a good floor, the use of one-inch mesh nettings over all openings and regularly closed doors at night.

CONVENIENCE

It is very often true that the novice, even though he has purchased a farm for the special purpose of going into the poultry business, will give less consideration to the location of his poultry houses, than to any other buildings. It is very likely that he will select his locations for his other buildings first and then place his poultry houses on what other convenient spaces are left available. Of course, the business-like method would be to locate the poultry houses first, and then to give the other buildings what locations were left.

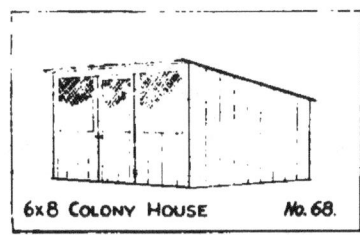
6x8 Colony House No. 68.

The question of labor is an item of great importance in the care of any livestock. The class that requires the closest attention to detail as a rule requires the most labor. Poultry keeping is wholly a business of details, consequently economy in labor is of prime importance. Buildings inconveniently located become expensive on account of unnecessary labor.

As poultry houses have to be visited several times each day every day in the year, it is more necessary to have them located conveniently than any other buildings on the farm. Operations must be performed frequently, so that inconveniences will cause not only extra expense in the care, but most likely neglect of operations that ought to be performed carefully each day.

All houses should be planned and located so that the attendant may need to do as little walking as possible. The house should be high enough to work in without inconvenience. There should be few partitions and fittings, and the latter should be movable and light in weight. Doors and gates should be self closing and should be large enough to allow the passage of a wheelbarrow. In the long houses containing several pens there should be a car suspended on a track, so that it may be pushed from one end of the house to the other, the doors between the pens being hung on double spring hinges to allow the passage of the car either way.

Where it is possible, running water should be in every pen and house, so that none need be carried. The so-called drip system is the best in which a faucet is adjusted to drop water into the receptacle at just about the rate the birds consume it. In the drip system it is much better to have the pipe connected to a separate supply, a barrel or tank, rather than to the main high pressure, because the drip cocks work much more accurately if they can be opened wider. Under heavy pressure the cocks have to be practically closed so there is only a very minute opening which is quite likely to clog frequently. Of course in very cold climates it is not practical to have this system in use in the winter weather.

All grain and feed should be kept as conveniently to the hens as possible to save steps. Also, where possible, the cellar for sprouting oats or storing vegetables should be made a part of the building or at least be handy to it. One can well afford to spend a little thought and money in making everything as convenient as possible so to make the work of caring for the flocks a pleasure instead of a drudgery.

LOCATION

As the laying house is the only house that is likely to have a permanent location on the modern poultry farm, we need consider only that under this head. Considerable discretion should be used in selecting a permanent location for any hen house. The laying house should have an exposure a little east of south. (One hour of morning sun will do more to keep the birds healthy and happy than two hours of afternoon sun.) It should be on high and dry land as intimated elsewhere. It should be located in a sheltered position, if such can be secured without special inconvenience. We would prefer land protected on the west by trees or hills. This is particularly important where the winters are cold and stormy. While fowl do not need warm houses, they must be protected from drafts or currents of air. Good air drainage should be sought so frosty air will be drained away. Houses of the popular open front construction are especially liable to be drafty in exposed locations.

It is desirable to locate the house where there may be shade for the birds when out in the yards. The ideal arrangement for this, of course, is to locate it in an orchard, as suggested in our chapter on poultry and fruit combination. Other shade, however, can be provided if an orchard is not desired. A selection should also be made, where land is available for yards on both sides of the house, provided yards are used. While the yards on the south side would be used the greater part of the time, it would be found very convenient to have the north yards to use on the hot days of summer, or at any time when the poultry man wishes to plough and plant the south yard.

VENTILATION

As ventilation depends to some extent on what kind of a location the house has, we may as well take up the matter here. In a house properly constructed and properly located, ventilation is not a difficult problem. There are various methods employed, such as houses having ventilation through the roof, foul air being carried off by pipes that come within two or three feet of the floor. There are also many mechanical devices, all more or less expensive, for giving scientific ventilation.

The general experience of poultry men, however, is in favor of open front or muslin front construction by which almost perfect circulation may be secured and the fowls at the same time kept in a good, hardy, vigorous condition. The open front and the muslin front differ only in that the former is open at all times, being inclosed only with wire netting. The curtain front is where there is a muslin curtain or shutter to put down over this opening in extremely cold, or stormy weather.

The muslin front house, however, makes possible a better control of conditions in the house, and may be considered imperative in cold climates. During most of the year, the muslin shutter is left wide open. Even when the muslin is closed, good ventilation is secured through it, and without draft. If, however, the muslin is wet, or heavily coated with dust, it will not allow any ventilation. Muslin shutters absorb and retain heat very much like glass and if closed when the sun is shining strongly, the house may become entirely too warm.

In the large 60 x 60 Twentieth Century houses that we employ on some of our plants we use a combination glass and muslin front. It will be noted that there is a possibility of a complete open front on the west, south and east sides; only the north side is closed tight. These open fronts may be closed either with glass windows that operate on hinges or on runs, or may be closed with muslin shutters. The judicious use of both the glass and muslin shutters enables the poultry man to have practically perfect conditions in his pen. The pens being so large require somewhat different equipment to handle the ventilation than the small pens. In ordinarily decent weather two full sides can be left open. In the summertime all three sides, but in case of a heavy wind or storm from any direction, the windward side of the house may be closed tightly or only with muslin, as preferred. While an opening on the south side of any hen house is satisfactory to leave open a great part of the time, most poultry-men realize that there are many disagreeable storms and winds that come from the south. On these days it will be found very convenient to be able to leave open either the east or west side, as may seem best.

POULTRY YARDS

As the poultry yards for all practical purposes are a part of the poultry house itself, it will not be out of place to say a few words regarding the same here. Where yards are used, they should be as large as possible, and usually it will be more satisfactory if they can be divided into two parts to permit alternate cropping, somewhat after the plan of rotation of chickens and crops, described in another chapter. The combination of north and south yards mentioned elsewhere in this chapter will also be found very convenient. The amount of space that should be allowed for each fowl in the yard depends on the nature of the soil. On an ordinarily good soil fifty square feet should be allowed for each bird. On land that does not drain off as well, or becomes more easily contaminated, much larger area is necessary. Great precaution must always be taken, however, that the land nearest the building, which is most occupied by the birds, does not become foul. Regardless of the size of the yard this is liable to occur, and about the only way to prevent it, is to have the division in the yard mentioned, so that at least one half of the yard can be devoted to crops each year. We use no yards in connection with our Twentieth Century houses.

Fences enclosing yards should be about six feet high, and should be so constructed that they may be readily taken down. There should be gates in each fence sufficiently large to permit a team to drive through.

NESTS

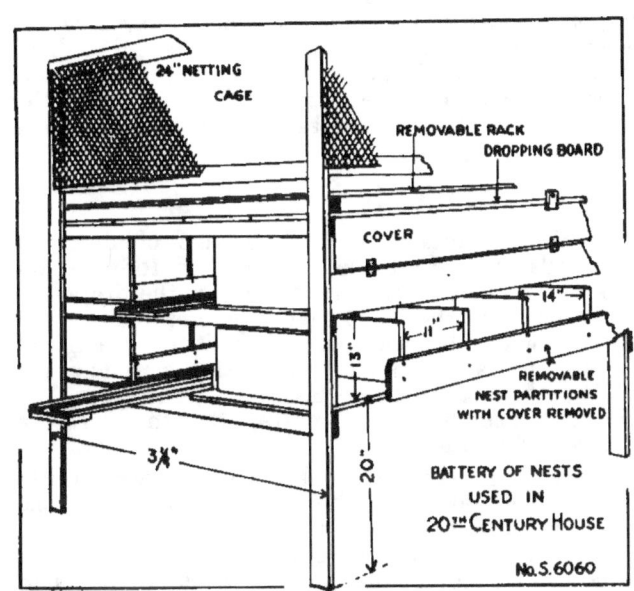

The nests in all laying houses should be so located that the interior of the nest will be dark. Hens are very retiring in their nature at the time they are ready to lay their eggs. They wish to be undisturbed, and if light, open nests are used, it will be found that the hen that is trying to lay will be continually disturbed by the other hens in the pen. Furthermore open nests encourage egg eating, which is a very unfortunate habit for one to allow hens to get into. In small houses the nests may be placed under the dropping board, but this should be avoided in any kind of a house if there is any other place where the nests can be located, for it is difficult to keep the nests under the dropping board free from lice and mites, and, also, the location of the nests there obstructs the floor space so that the actual number of square feet per bird that is available is cut down materially, because the floor under the dropping boards is so cut off by the nests that it is of no practical use. In houses where there are small pens, the nests may be made a part of the partition, being half in one pen, and half in the next. In large houses, like the Twentieth Century, it is much more desirable that the nests be made in batteries and stand either at the sides, or in the middle of the floor.

Experiments with different types and arrangements of nests at our Westboro plant have shown to our satisfaction that nests arranged in batteries, located away from the walls, are much preferable in the Twentieth Century houses at least. Nests located against the wall have a tendency to get damp, also they interfere with the light and air from the windows. At Westboro we have discarded the wall nests entirely and use batteries exclusively. The hens themselves show a very great preference for the batteries, and in most cases they may be trusted to know what is best for them.

Nests should be so constructed that they can be readily cleaned and thoroughly disinfected. If they are removable so that they can be taken out of doors, so much the better.

Nests should also have plenty of room to approach them on two or three sides. It is a well known fact that some hens in seeking a nest will always drive off other hens, no matter how many vacant nests may be available. If the nest is so arranged that it can be approached from only one side, when one hen is driving another from a nest there is likely to be a combat, resulting in broken eggs. This leads to egg eating. Nests should also be constructed with partitions sufficiently high so that it will be impossible for a hen to draw eggs from one nest to another and care should be exercised that the nests are not so filled with nest material that the hens can do this.

The rule as to the number of nests necessary in a pen varies according to the size of the pen. In large pens there should be one nest to five fowls, and in very small pens one nest to four fowls will probably be found sufficient.

STRAW ROOFS

Elsewhere in this same chapter we made a suggestion regarding straw roofs. We do not mean by this that the roof is made of straw, but in the ordinary pitched roofed building, instead of leaving everything open to the roof, or instead of putting in a tightly sealed wooden ceiling, we make a ceiling of boards placed three or four inches apart on the cross timbers. This ceiling is best described as a slatted ceiling. On the upper side of this ceiling we place two or three feet of straw. The object of this ceiling is to prevent too much cold, heat, or dampness coming down from the roof, without sacrificing the advantage of the ventilation of the large roof. The straw affords perfect ventilation, the foul air from the pen below percolating slowly up through the straw and passing off through ventilators in the roof. Furthermore, this sort of ceiling has the advantage that all ceilings have, of lessening the number of cubic feet that the birds have to keep warm in the winter. Low pens will always be found to be much more comfortable than high, airy pens, and the result may be accomplished by the straw roof without at the same time preventing the escape of the foul air.

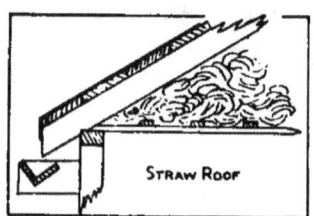

DROPPING BOARDS

There are different opinions as to the advantage or disadvantage of dropping boards. The advocates of dropping boards argue that they are needed for sanitary purposes and that they pay for themselves in the proper preservation of the droppings for fertilizer, keeping them free from the litter on the floor; that it is more convenient to remove the droppings from dropping boards and that the boards make it much easier to keep the houses clean. All of these arguments are worthy of careful consideration. Other poultry men, however, claim that they cause more unsanitary conditions than the lack of them, that they are a harboring place for lice and mites, that they waste a great deal of room and that they are a cause of drafts.

As far as sanitation goes, it is quite feasible to have the perches near the floor and to allow the droppings to drop on the floor. A partition should be built, however, about one foot high, to separate this part of the floor from that part which is covered with litter, and no litter permitted to be under the perches. The floor beneath the perches should be covered with a good absorbent, like good dry loam or some of the especially prepared absorbents of which O. K. litter is a good example. A thin layer of this absorbent should be spread over the droppings each day. As to the value of the droppings accumulated in this way we believe that it will be found that if they are properly kept, they will be much more valuable than those that are accumulated on the dropping boards, for the reason that the absorbent with which they are mixed prevents the escape of the very volatile ammonia. The chief difficulty, however, is that sufficient care may not be exercised to keep them free from the litter or to spread on the absorbent.

The dropping boards possibly insure best conditions in the ordinary hen house, but they have no advantage over the other arrangement when the other arrangement is properly carried out. Where dropping boards are used they should be sufficiently far from the perches to admit of easy cleaning.

DROP CURTAINS

In our 60 x 60 houses we do not use drop curtains in front of the birds for the reason that the openings may be so well handled with respect to the weather, and because in these houses the birds are so far away from the openings, that they cannot possibly get any drafts; practice shows that we need no curtains. In the case of houses that are very narrow, in which the birds at roost are not more than ten feet away from the curtains that are in front, there is no need of further curtains in front of the roosts.

There are houses, however, the width of which is in between these two, where curtains have been found to be very desirable. We speak of houses from fifteen to twenty-five feet wide, like for example, the Gowell laying house, originated at Orono, Maine. It will be found essential to have a muslin curtain in front of the birds in such houses, but it will be found advisable also to supply ventilation to the roosting place supplementary to that which goes through the muslin curtains. This is provided for by having openings of greater or less size, as the circumstances seem to require, above the curtain frames. In the use of drop curtains on the roosts great care must be exercised that the birds are not confined so closely that they sweat. If they do become overheated, when they get down on the cold floor in the morning, they will be quite likely to catch cold, which cold will spread through the flock with consequent curtailment of egg production.

Neither should the poultry man carry the matter too far the other way, and fail to put the curtains down at night when the weather requires it. One must watch the condition of his birds very carefully during the fall when the nights gradually become colder. It is a serious matter from the standpoint of production, to allow the birds to be "Frozen up." It does not pay to try experiments on the birds in such matters as seeing how long they can get along without the curtains closed. The curtains should be two feet in front of the roost.

PRACTICAL HOUSES

As we are trying to point out "One Road to Poultry Success" we will confine ourselves, as far as possible, to single types of houses which we know from our experience are suitable for the different stages of the chicken's growth. There is a general impression in the minds of some beginners that any house is good enough to house poultry that will keep out the weather, but all beginners should take it for granted that what we have said above in this chapter we know to be true from our own experience. On the other hand it is not always necessary to build entirely new houses. Many times the old farm will have buildings that may be well utilized if simple changes are made in them or, may be, if the location of the house is changed. In what follows we shall give only the most important details. We expect the builder to exercise some ingenuity in planning his details, but we do not recommend the novice to try many experiments in developing his own new ideas before having the advice of some successful poultry raiser. For those that wish to get every detail of construction of the houses we shall describe, we have prepared a series of blueprints as enumerated in the advertising section, and which we will send postpaid to anyone wanting them for the nominal price of fifty cents each. As we describe each building, we will give also, the number of the prints showing the details.

BROODING HOUSES

In our experience we have tried a great many different types of brooders and brooding houses. We are familiar from experience with the difficulties the poultry man has to encounter in trying to raise chickens with the old fashioned lamp heated brooders. We discarded the lamp brooders for a number of years of experience with different types of hot water systems. Any of these were so much of an improvement over the old lamp brooder system that we felt as if the millenium in brooding had come. A few seasons, however, convinced us that even these were not perfect and that something better would be discovered in time.

COLONY BROODERS

During the past few years we have been trying out various types of Colony brooders as they have been put on the market. The fact that we have now discarded all systems of brooding in favor of the coal-burning Colony type is sufficient proof that we consider this the best system that is available today.

Under the system of long, pipe brooders, chickens were kept in flocks of fifty to a hundred, under hovers two and one half to three feet in diameter. The height of these hovers from the floor was about eight inches. In front of each hover was a run of varying size, but usually about three feet wide and seven or eight feet long. Supplementary to this was an outdoor run of greater or less dimensions.

The faults of this system were many. In the first place, flocks of the size mentioned were very much crowded in the space available under the hover. If smaller flocks were used a longer system was required, necessitating in turn more investment. There was a great lack of both room and ventilation. The great numbers of separate flocks required by any great numbers of chicks meant an interminable amount of work, with the result that work that should be done was not done, or at least was not done on time, so that the chicks very frequently were living in unsanitary conditions. Then another objection was the fact that the chicks had to be raised on the same land, out of doors, year after year, which can be very rarely done successfully. And a still further objection to the whole system was the amount of money required to install it.

ADVANTAGES OF NEW SYSTEM

The coal-burning Colony brooder system overcomes all of these disadvantages. The chicks have not only much more room, as a flock under the hover, but each chick, as a single chick, has a greatly increased area to run around in. The air under the hover is not only greater in volume by many times, but it is also much purer, the foul air being carried off automatically. There is a still further advantage, that the poultry man can provide all the space that the chicks require in the pen outside of the hover with a very moderate investment, and the whole brooder, colony house and all, can be moved to new land each year, or as many times each year, as is desired. And there is the very great advantage of this system, that the original cost of installing it, as compared with the old systems, is agreeably small.

The convenience to the poultry man of handling the Colony brooder system is probably the most appreciated advantage of all. His chicks do better, not only because they have better regulated heat, more room, more air, and better sanitation, but also because the poultry man has more time to be among his flocks watching their progress.

Since the time that the first coal-burning colony brooder was introduced there have been many types on the market offering many improvements, over the original product. The novice, however, should take considerable pains in selecting the type that he will use, because many of the types on the market now are not all that they should be, especially in the matter of regulation, convenience of handling and coal capacity. Furthermore, there seems to be a contest between brooder manufacturers as regards claims of big capacities for their brooders. The question for the novice to decide from reading and from investigation, is as to which brooder will best brood the chicks regardless of the number of chicks, and then to decide from experience how many that brooder will handle with good results. The size of flock that it will handle will depend about as much on the size of the brooder room and the size of the birds to be brooded, as on the size of the brooder itself. He must remember that the chicks grow to several times their size at hatching time before they can be removed from the brooder.

RULE FOR SIZE

If the brooder house is too small, there will not be room to work during the day perhaps for the number of chicks that the brooder will take care of under other conditions. It is impossible, however, to lay down any hard and fast rule as to the number of chicks to put into a brooder. As a general rule we would suggest that an 8 x 10 Colony house should not have over two hundred chicks, a 12 x 12 not over three hundred, while a 14 x 14 would easily take care of four hundred, and in most cases five hundred.

Regarding the colony houses themselves, which are to be used as combination brooders and range houses, the 8 x 10 colony house (Print No. 810) is the largest house that can be moved conveniently to any part of the farm. This house should be equipped with one of the smaller types of brooders. It is also the smallest house that should be used in this colony brooder system. It can be used later for ranging about eighty chicks to maturity.

The 14 x 14 colony house (Print No. 1414) will raise a greater percentage of chicks than will any smaller size. The area in this house is so large that the corners are sufficiently inconvenient to keep the chicks from crowding into them, and there is enough variation of temperature so that the chicks have the necessary changes. The house is too large to be moved very great distances, but it may be moved short distances each year, say fifty feet, so that the chicks may have new grounds. In this connection we recommend strongly that a small piece of ground be fenced off in reserve each year for the following year's use, as a new location for the house at the beginning of that season: about one or two square feet for each chick should be allowed for this purpose. If this idea is followed nine-tenths of the usual bowel trouble will be eliminated. The 14 x 14 house is one that is continually useful. It can be used, not only as a colony brooder house, but also will accommodate about one hundred and fifty growing chicks to maturity and may then be used for wintering thirty-six to fifty layers.

LONG BROODER HOUSES

The suggestions above regarding size and capacity hold good for long permanent brooder houses. For the long house with many pens is nothing more or less than so many colony houses put together. Pens arranged in this way, in a long house, may be equipped with colony brooders like the separate houses; in fact we have such a house so equipped on our Holliston plant and the expense of labor in handling chicks in this way represents considerable saving. There are many disadvantages, however, to having the pens grouped together. There is one particularly big disadvantage, which is that the house cannot be moved to fresh ground each year, and cannot be used during the season to range stock.

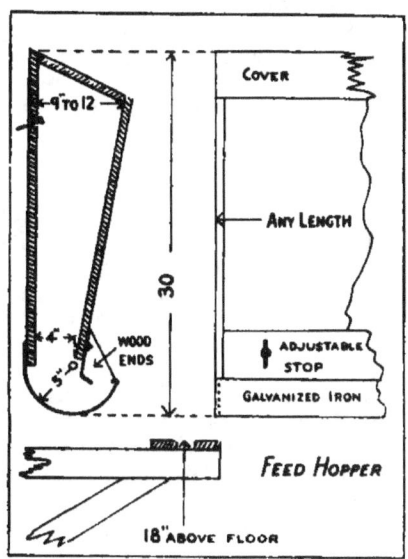

Although we have such houses on our plants we wish to discourage the use of them so far as we can. We shall continue to use them, because we cannot afford to give them up, as yet. We are, however, making all our increases of capacity by means of the colony brooder system, installed in separate movable houses. If one has a sandy soil and will be satisfied to use the long brooder house for only one lot of chicks each year, it can be made to give fairly good results, but unless one's plant already includes such a house, we should advise strongly against building that type. We have, however, drawings showing all the details of the best model of this type of house (Print No. 140.)

HEN BROODING

For small numbers of chicks, or for chicks of special matings, where a little extra cost is not prohibitive, the hen is still the best brooder. A 4 x 8 house (Print No. 48) is well adapted to caring for two hens with their broods and the drawings show how to arrange the house for such purpose. In hen brooding fifteen to twenty-five incubator chicks may be given to each hen. If there are two varieties of chicks, care must be taken to give each hen some of each kind; otherwise, when the chicks get accidently mixed, the hens may kill the intruder.

In cases of pedigreed and exhibition stock, where it is necessary to keep an exact detailed account of the breeding of each bird, and to watch the growth and development of each bird, it is absolutely necessary to brood by hens and practically necessary to hatch by hens. Different eggs may be kept separate in the compartment of an incubator, but in spite of the greatest care, the chicks occasionally get mixed, so that it is safer to use the hen for incubating also.

There is an old fashioned notion that the hen will hatch more and better chicks than the incubator. We can speak only from our own experience, which teaches us however, that the incubator is more reliable in both respects than the hen, and that the work can be carried along with very much less expense and bother. With regard to brooding, our views are entirely different. We have yet to discover a brooder that will give as good results as the hen, and we believe that the hen should be used on all special chicks. For other details regarding handling of chicks in hen brooding, see the chapter on "Feeding."

RANGE HOUSES

Where the colony system of brooding is used, the same colony houses will be used for the chicks on range, up to their capacity. With ordinarily good luck, however, extra colony houses will be needed. The cheapest house to use for this purpose is the 4'x 8 house (Print No. 48) mentioned above. This house will range forty chicks to maturity. It can be moved by one man on a one horse drag to any part of the farm, and can be placed under a shade tree, or in the open, as one desires. From the standpoint of economy of construction, it has no equal. At one time, we let a contract for one hundred of these houses, complete, for $4.00 each. During the present unusual times, however, the cost is probably forty to fifty per cent higher. The poultry man who is his own builder should be able to save something on a contract price. The 4 x 8 house, however, is somewhat too small for use in all places, that is, it is not sufficiently deep to afford ample protection in all locations. For locations well protected by trees, etc. from beating storms, it is very satisfactory. Our 6 x 8 house (Print No. 68) however, allows more opportunity for the birds to get back from the weather. We consider the 6 x 8 house the most satisfactory range house that can be built for a reasonable cost.

If shade trees are not available on the range selected, some sort of shade should be provided, otherwise the young growing chicks may become stunted by the extreme heat of the middle summer. A strip of corn thirty-five feet wide and as long as necessary will furnish the best of shade. The 4 x 8 colony houses may be placed on both sides of the strip about seventy five or one hundred feet apart, or the 8 x 10 houses one hundred to one hundred and twenty-five feet apart. Further particulars regarding the handling of chicks on range will also be found in the chapter on "Feeding."

LAYING HOUSES

There are innumerable different types of laying houses of different degrees of desirability or undesirability. Nine-tenths of the new models of houses, or changes in old models, that are offered to the public by would-be poultry benefactors are undesirable and uncalled for. The plain ordinary poultry house of the ordinary dimensions and old style, provided it is well located, is kept sanitary and dry, and has an open front, or muslin front, is better than any of these new designs with very few exceptions.

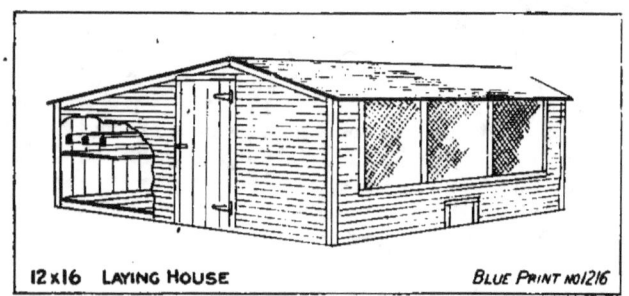

There is a class of people that are never satisfied unless trying some new wrinkle. They are honest enough in their purpose and in their conviction, although perhaps misled by unusual circumstances, but they are not really helpful to the business. Then again there is a class of people who get up new designs for the sake of the profit they can get from selling the plans or from the advertising it gives them personally. Some of these designs have the same advantage as a certain cold remedy we once read about, that "wouldn't do any harm, if it didn't do any good." But we fail to see any good reason for changing from a design that has been proven by experience to be good, to one that makes no claim to being any better.

There are many advocates of the laying houses with the monitor, or semi-monitor roofs, and many of these advocates are men prominent in the business. We cannot, however, see any advantage in them, and in fact consider them very unsatisfactory, except in warm weather. They are cold at night and do not allow sufficient sun to enter on the birds in the daytime.

LAYING HOUSE AXIOMS

There are two primary facts that one may keep in mind in designing a hen house.

First: A square house economizes in lumber.
Second: A large house economizes in labor.

Other things to keep in mind are that alley ways should be avoided. It is better to give the space to the stock, and alley ways are inclined to cause the poultry man to put less time in among his birds.

Low houses are best. There should be just enough room to stand or walk under the straw or wooden ceiling.

The amount of floor space allowed per bird in large houses should be three and one half to four square feet; in small houses four to five square feet.

TWENTIETH CENTURY HOUSES

Among the new models of houses that have been advocated in recent years is the so-called Twentieth Century house, which we know from our own experience is practical in design, economical in both investment and labor, and highly satisfactory from an operating standpoint. The credit for designing this house, we believe, properly belongs to Mr. Carl Aamodt, who at one time had charge of one of our plants and who made the design while in our employ.

The details of this house are shown in three plans (Print No. 6060, Parts 1-2-3.) This house is sixty feet each way, has a straw roof and accommodates one thousand layers of American breed, or twelve hundred layers of any Mediterranean breed. We have three of these houses on our Westboro farm. There have been several others built by other breeders upon our recommendation. The results in every case from the standpoint of health of the birds, egg production and profit earned, have been extremely gratifying.

The extraordinary fact about them is, that the fertility from breeders housed in them in the large flocks, no matter what the breed, is greater than from breeders housed in small pens. The egg yield is also better. So with labor reduced to a minimum, as it is in these houses, it follows that the profit is the maximum. From the standpoint of investment the house is also profitable. It can be built at the rate of about $1.25 per layer: many of them have been built for much less than this, but even the higher figure shows a considerable saving over the rate at which any other type of house can be built.

Over the straw ceiling are located grain bins, in which may be kept a week's or a month's supply of grain, as one chooses. The grain is delivered to the feeder on the floor below by means of spouts, so that the feeding of one thousand hens is a matter of only a few moments work. Surplus straw may also be kept in the loft above, and, if one desires, a cellar may be built under the house, in which to store all vegetables, and which cellar may also serve as an incubator cellar. This makes a very compact, complete equipment. We know of nothing that can approach it in economy or results. An illustration on another page will give the reader a very fair idea of this building.

TWENTIETH CENTURY JR.

For those who expect to limit themselves to five hundred layers we have plans for a house similar to the Twentieth Century, which we call the Twentieth Century Junior. This is a smaller house, and will accommodate five hundred layers. It is also shown in three drawings. (Print No. 4248, Parts 1-2-3.)

12 X 16

For less than five hundred layers a long house with 12 x 16 pens gives excellent results. The details of this are shown in Print No. 1216. Each pen will accommodate from thirty-six to fifty hens. If the poultry man prefers, he may keep the hens (and perhaps to better advantage) in colony houses, 12 x 16, each house being, of course, a facsimile of the 12 x 16 pen in the long house. Or the 14 x 14 colony, as we have said elsewhere, makes an excellent laying house for thirty-six to fifty hens.

For smaller lots of hens for the city back-lot, 8 x 10 houses will be found ample in size and highly satisfactory in character. We show these in Print No. 810.

INCUBATOR CELLARS

For somewhat the same reason that it does not pay to keep hens in too small numbers the small breeder, especially the novice, will do well not to consider running an incubator unless he can handle one of considerable capacity (See Chapter on "General Topics.") If, however, it seems best to him to have an incubator cellar, we would suggest that he obtain from the Hall Mammoth Incubator Company of Little Falls, New York, one of their catalogs which gives very fine detailed drawings of a satisfactory cellar.

THINGS TO AVOID

We remember at one time reading in one of the Cyphers Company's bulletins, a chapter entitled "Things to Avoid" in construction of poultry houses. This article struck us very forcibly and it covered many things worth considering. We will try to give these as far as possible from memory.

1. —AVOID TOO MUCH OR TOO LITTLE GLASS. One will make the house too warm, the other too dark. Glass will draw the sun's heat in the daytime, causing the pen to warm up very materially, but at night, when the source of heat is removed, the pen will become cold again, and just as

cold as if it had not been warm in the daytime. The great change in temperature is very bad for the birds. A dark house is very frequently damp, and is so unattractive, that the hens will stay out of doors rather than in the pen, regardless of the weather.

2.—**AVOID FACING THE HOUSE AWAY FROM THE SUNLIGHT.** Houses should be faced to the south or southeast, or in whatever direction they will receive the most sunlight and avoid prevailing winds from blowing directly into them.

3.—**CROWDED HOUSES.** Never overcrowd a hen-pen. "Cut your garment according to your cloth." If you have proper room for only one dozen hens, do not try to keep more than one dozen. Smaller flocks having ample room will give better returns than large flocks in a crowded pen.

4.—**AVOID LOW FRONTS.** The type of houses, which, as the Irishman said, have the back in the front, are not satisfactory. We mean those houses where the low side is used for the front. They are neither as well lighted nor as well aired as when the low side is in the back. The roof sloping to the south also makes the house warmer in the summer, and does not help any in the winter.

5.—**AVOID NARROW HOUSES, OR WIDE HOUSES.** A long, narrow house reflects the outside changes of temperature too easily. Roosts are too near the opening in narrow houses. A wide house costs more to build.

6.—**AVOID HIGH CEILINGS.** We have given reasons for this in another part of this chapter.

7.—**AVOID TOO MUCH OR TOO LITTLE VENTILATION.** Where there is too much ventilation the pen is drafty and cold in winter. Where there is too little, there is dampness and sickness, with the consequent lower egg production and high mortality.

8.—**AVOID "FREEZING" UP THE HENS.** Houses in which combs are frosted are either not properly built, or are not properly cared for. One cannot expect his birds to give results in an icehouse.

9.—**AVOID WARM HOUSES.** We have shown why elsewhere.

10.—**AVOID DAMPNESS.**

11.—**AVOID UNNECESSARY EXPENSE AND UNWISE ECONOMY.** The house should be built as good as need be to accomplish the work that it was intended for, in good shape. Do not create too much interest charge to pay, nor build the house so cheaply that the fowls are uncomfortable and the place too inconvenient to work in.

12.—**AVOID POOR MATERIALS AND POOR WORKMANSHIP.** A poultry house may be plain, with no unnecessary frills, but it does not follow that it should not be neatly and carefully built of good, substantial material.

CHAPTER FIVE

FEEDING AND CARE OF CHICKS FROM SHELL TO MATURITY AND BEYOND

(A reprint of a bulletin published by Pittsfield Poultry Farms that has had a circulation of nearly 100,000. This Bulletin will be sent free to those addressing the Company and enclosing a stamp for postage.)

This method is not the only satisfactory method of feeding but is one that has given the best of satisfaction on hundreds of farms.

OBJECT

The object of this folder is to give the inexperienced some definite concrete information which will be of real value. We will suggest nothing but what we believe is of utmost importance. This folder is mailed to all our Baby Chick customers three weeks before we make shipment. We have at that time completed our sale, so the only axe we have to grind, is to see that our chicks get the best of care.

IN GENERAL

It is a very nice matter to determine when the chicks have had enough to keep them growing properly and still not enough to clog their desire for food.

It is quite apparent that the one feeding must be a close observer and must assume that something is wrong all the time.

For baby chicks, temperature and care are more important than food. If you use the large brooder system, you must use every ingenuity to keep the different hovers properly heated. (We have taken out all our extensive hot water systems.) The chicks must also be kept dry, clean, busy and slightly hungry, and every day must show a little growth over the day before.

You must not make any change in food, except you do it very gradually. You should examine all your grains carefully before feeding. Musty grain is the cause of (we might say) most chick losses. If you feed a dry mash, see that they have it fresh, and see that it is before them at all times. If you feed a moist mash it must be crumbly, and the troughs must be thoroughly cleaned after each feeding.

Regular and sanitary methods cost almost nothing, but they are of the utmost value. The chick, up to the time it is ready to lay, should not be forced, yet it should grow slowly and surely every day. The common mistake is to give the chicks too much while in the brooder, and too little after they are on range. The chick on range should get a lot of exercise to develop its frame. It requires food not only to maintain life, but also to keep it growing. You should handle your birds every day to see that they have a full crop at night, and to see that they are growing. If you don't, some day in the Fall you will handle them to find that you have little else than feathers.

SIZE OF GRAINS

You must feed small grains to chicks. For breeders of course, larger grains are used, but the finer the scratch grain, the more exercise they will get and the better will be the fertility. Some ground grains are very desirable, for layers that are to be forced for eggs.

EXERCISE

Exercise is especially necessary for growing birds and for breeders, but for forced layers it is less desirable except just enough to keep them healthy and to prevent them from getting fat. A heavy layer will eat a lot without growing fat, because she not only requires the same amount as any hen to maintain life, but also uses up a great additional amount of nourishment in compensating for the drain on her system in producing eggs.

The secret of heavy layers is about half properly bred parents and a good big half skillful feeding and care.

COMPOSITION OF FOOD

In order that the feeder may change his food intelligently to meet the constantly changing conditions, it is desirable to know something about the digestible nutrients in the food.

Protein, which we will designate as (P), is the name applied to the nitrogenous compounds. Protein is necessary to develop and renew the body and to supply the material for the white of the egg, which in turn supplies the material for the baby chick. So that Protein in fair quantities is quite essential to the heavy layer.

Carbohydrates and Fats which we will designate as (C) and (F) supply the energy for the body and the stored up fat. It also supplies the material for the yolk of the egg, which later is found enveloped in the chick's stomach and supplies the energy to the chick while it is developing in the shell, and for about three days after it has left the shell. The presence of the yolk in the developed chick's stomach accounts for the possibility of long shipments of day-old chicks.

Minerals are also included in the food nutrients, but are usually fed separately as the grains contain only small amounts of mineral matter. Calcium and Phosphate are the principal minerals necessary for the body, and the egg shell. If the birds do not get enough you will notice it in the shell. The usual practice is to place an abundant supply before them at all times and let them help themselves.

The "Ration" is the amount of feed of all kinds, given a bird in a day.

> 1st to maintain its body.
> 2nd to produce its growth.
> 3rd to produce energy.
> 4th produce its by-products;—its surplus fat, its egg and new feathers.

As these four requirements continually change so must the Ration continually change. It is approximately true that the food given to a laying hen is utilized as follows:

> One-third is used to maintain its body, to produce growth, and to produce energy.
> One-third to produce eggs.
> One-third goes to waste.

The Nutrient Ratio in any grain is the ratio between its P. and its C. F. That is, 1 to 4 would mean 1 part of P. to 4 parts of C. and F. combined. You can easily understand that the ratio must change as conditions change. 1 to 5 or 1 to 6 may be a proper ratio for birds during certain stages of their growth, but 1 to 4.5 may be better for a heavy laying bird. The ratios of feeds given later are about what we believe will suit average conditions, but they may have to be changed to fit circumstances. To produce the growth of a pullet requires practically the same amount of P. as it requires of C. and F. In fact it requires a little more of P. The maintenance of the body and the production of the egg also require slightly more of P.

For producing energy, however, C. and F. only enter, and energy is what we have to produce most of.

A growing chick, for example, requires a feed in which the nutrient ratio is high, perhaps as high as 1 to 5 or 1 to 6. Simply for the growth and maintenance of the body alone, the ratio would be something like 5 to 4, but the actual quantity of food required for the body is very small compared with the food required to produce energy, and as energy is derived wholly from C. and F., the required ratio is high.

If the chicks are given sufficient exercise to develop a rugged body, most of the food goes into energy, and just as soon as you restrict the exercise, some of the food which went into energy goes into fat. It is a matter requiring very careful attention, this regulation of food and exercise.

ONE ROAD TO POULTRY SUCCESS

If you started with a balanced ration for a bird just ready to lay, theoretically you should add for the eggs, approximately equal quantity of P. and equal quantity of C. F., because the eggs require about equal quantities.

When a bird is in heavy laying, the ratio approaches 1 to 4. Of course the ratio is made up of the whole days feed, and as it is inconvenient to change the ratio of the scratch feed, most of the change has to be made in the mash. The skillful feeder must change the feed one way or the other as he notices undesired changes in the birds.

IN OTHER WORDS

A growing chick should have a food rich in fat producing elements but should have sufficient exercise so that it does not produce too much fat.

A bird that has its growth should have a less fattening food, or one with a larger percentage of protein.

When the bird commences to lay it should have an additional supply of food and this addition should be about equal parts of fat producing elements and Protein, and this additional amount should be increased in proportion to the egg yield.

No rule can be laid down for the exact amounts of food to be given, nor, for the nutrient ratio of the food, but by handling the birds every day, one can easily determine if a change is needed and what change.

In order to give an idea how to make the changes, we give the qualities of some of the more common grains.

Wheat, Corn, Barley and Oats are all rich in C. F., and consequently are useful for growing stock and to fatten stock. They do not, however, contain enough P. to supply the needs of a growing bird, and much less the needs of a heavy layer.

Beef Scrap, Fish Meal, Oil Meal, Gluten and Bone contain very much larger percentage of P. and consequently are necessary to balance the ration.

Oats would be perhaps the best one feed if it were not for the waste of the hull.

Wheat is really the best one grain.

Corn is popular on account of its price, and popular with the birds on account of its color, taste and size.

The main advantage of using several grains is to give variety to the birds. Birds, however, could be fed on wheat and beef scraps for some time without ill effects.

Oil Meal is fed in small quantities for the plumage.

Alfalfa, Clover, Sprouted Oats, etc., supply P. C. and F. to the birds, but the real reason for feeding them is that they keep the digestive organs in proper working order. Sprouted oats are especially valuable for breeders, in so much as they help the fertility of the egg. Any green food when cured, should be cured as green as possible.

Onions. Small quantities of onions we have found very helpful to layers. It seems to be a great relish, and if fed in small quantities as a condiment twice a month, is a good tonic. Onion seconds may be bought usually for about $1.00 per bag.

Milk, either fed in fountains or in the mash, is very helpful. The Lactic acid in sour milk aids digestion and prolongs the life of the hen. As a first feed for chicks it prepares the digestive organs to receive food.

Oyster Shells, or a substitute, are absolutely necessary.

Charcoal, we believe, keeps the blood in pure condition and aids digestion. However, some people profess not to believe it. But as it does no harm, it may as well be fed.

Mustard is said to help laying hens, and should be fed in the mash. One heaping teaspoonful to a dozen birds is said to be the proper amount. We have not as yet made up our minds whether it pays or not. But we believe that condiments are helpful to housed birds.

Potassium Permanganate. A few crystals placed in the drinking water, enough to give it a good red color, is an excellent preventive and disinfectant. It is not a medicine.

Moist Mash. For broilers and layers we believe in the moist mash. Especially for forced layers. Our observation indicates that the birds get all they need in a short time and are not hanging around the dry mash all day, but are actively out hunting in the litter or in the yard. Of course grain should be fed in the yard when conditions permit.

By-Produce. Often stale bread, crackers, etc., can be bought at a price that makes them look attractive as a feed. Broken sweet crackers are especially good for fattening birds.

Fish Scrap. In the past we have not had very satisfactory results from the use of Fish Scrap, chiefly because we were unable to get a product that was uniformly good. We found that much of the scrap was either oily or tainted and caused bowel troubles. Also if used in quantity it imparted a fishy taste to the eggs. At the present moment we have convinced ourselves that we have found in Chic-Chuk (a fish meal) a very desirable substitute for Beef Scrap and we have placed a large order for our 1917 requirements.

METHODS

There are probably a hundred, yes, a thousand, methods of feeding that are equally good, provided, of course, they are used consistently, that the chicks are not fed one method and the layers by a radically different method. We are going to describe our method as one of these possible thousand, and if you have not a method satisfactory to yourself, you can consider it. Remember if you do not start it from the chick up, you should change to it very gradually, for a rapid change in method often leads to very harsh results.

BROODERS

About a day before you expect your chicks, you should start up your brooder, and cover the whole floor with sand the depth of one inch, and the day you put the chicks in, add one-half inch of fine cut litter, such as alfalfa or straw. This litter can be added to as chicks grow older, and as foul litter is removed.

The droppings under the brooder should be cleaned out every day, and cleaned out from the whole pen at least twice a week. We prefer the colony type of coal-burning brooder, such as is advertised at our invitation, in the advertising section of this book.

FIRST FEED

In what follows do not mistake the first feed to mean first day hatched. The first feed should be given whenever the chicks seem anxious to eat, usually about thirty-six hours after they are hatched, although it may be delayed to forty-eight hours. They have the yolk of the egg in their stomach only partly digested, and should have time to pick up some grit to get their crops and gizzards into proper working order.

FIRST FEEDING DAY (A)

1st Feed.—Sour milk, seeing that each chick gets some of it. It should not be fed except this once, unless it can be fed regularly. (About one quart to five hundred chicks.)

2nd Feed.—Cut oatmeal in the litter with a little grit mixed in.

3rd Feed.—Commercial Chick Food. (We use Wirthmore's Chick Feed.) Temperature 100 degrees, two inches from the floor under hover. The first three or four days the chicks should be fenced in near the hover, to teach them to go there when cold. With a colony brooder they should have only about one foot all around the outside of hover.

SECOND AND THIRD DAY (B)

1st Feed.—Cut oatmeal in litter.

2nd to 5th Feed.—Four times during the balance of the day, chick feed in litter. Keep the chicks slightly hungry, and cut down the feed if you find unconsumed feed in litter. Care is more important than feed the first week. Temperature 95° to 100°. You cannot go entirely by the thermometer. You should go by your own feelings and by the action of the chicks. The temperature should be high enough so that at any time of the day you can see a few chicks stretched out resting, with the appearance of being dead. If the chicks crowd, you have either too many under hover, or the heat is too low. With a colony type of brooder there should be a zone about six inches all around the stove so hot that the chicks will seldom go there. At night after dark, see if they are all lying out flat, each one separate from the other. If they crowd at night, the next day they will crowd all day in trying to get some rest. Our observation tells us that a chick wants a nap every hour. It is only for a few moments at a time, but see that it can get it peacefully.

As suggested above, sour milk should not be fed after the first feed unless it can be fed every day. On most farms it is inconvenient to obtain a regular supply of sour milk and further many object to its use on account of the bother necessary to feed it right. We have found that a most excellent substitute is Blatchford's Milk Mash and we recommend that it be kept before the chicks in hoppers at all times. It will produce especially noticeable good results on your broilers. Don't feed sour milk after the first day, if you use the milk mash, as they do not work well together.

FOURTH TO TENTH DAY (C)

Same as (B) except feed at noon cut cabbage, sprouted oats or other green food, sparingly at first. We feed the top of sprouted oats to chicks (and the sprouts to older birds.) If you wish to give a variety you can feed a Johnny cake, moistened with sweet milk. Temperature 90° to 95°. Care still important. Look out for the droppings surely every day.

ELEVENTH TO TWENTY-EIGHTH DAY (D)

Chick feed four times during the day. Green food at 10 A. M. and 2 P. M. Dry mash should be fed in the morning, what they will clean up during the day. Formula— 100 cornmeal, 100 bran, 150 middlings, 50 fine ground oats, 50 fine beef scraps, 20 alfalfa meal. (All quantities are pounds.) The ground oats should be omitted if not fine and of best quality. Watch out for mustiness in the grains. Also when you let the chicks out, be sure the ground is sweet and be sure that there are no standing pools of water. Pools of water often cause a fifty per cent loss in chicks. Temperature on the twenty-first day usually about 85 degrees. Losses are apt to be highest from the tenth to twentieth day. If the chicks crowd in the corners of house, bank corners with straw.

FOURTH WEEK TO TWELFTH WEEK (E)

Same as (D) except gradually reduce proportion of Chick Feed and gradually change to coarser grains, so that at the end of the eighth week, chick feed has been wholly cut out. The scratch feed will then be 100 cracked corn, 100 cracked wheat, 50 hulled oats or pinhead oatmeal. Feed green food three times per day. Change dry mash to growing feed,— 100 cornmeal, 100 ground oats, 75 beef scrap, 100 bran, 150 middlings, 25 bone meal, 25 alfalfa meal, 10 charcoal, 1 salt, but remember to change very gradually.

FOR BROILERS

At the end of six weeks those chicks which are to be fattened for broilers should be separated from the other chicks, and placed in somewhat more confined quarters. Equal parts of cracked corn and wheat should be kept before them at all times in shallow boxes. Also sifted beef scrap should be fed in hoppers. Green food twice a day. A mash moistened with milk should be fed at 9 A. M. and 2 P. M. (What they will clean up in fifteen minutes.) If you have plenty of skim milk it should be kept before them. Mash formula— 50 bran, 150 middlings, 125 cornmeal, 100 ground oats, 100 beef scraps, 25 alfalfa meal, 10 charcoal. At the end of eight weeks, if conditions and weather are right, the broilers should be ready for market. If you put the broilers on range, give them plenty of shade, and have cracked corn handy so that they will not be inclined to wander far. Some breeds can not be separated for the fattening process before they are eight or ten weeks old.

THREE MONTHS TO MATURITY (F) (On Range)

Growing feed kept before them at all times, but gradually increase the corn [meal to 125 pounds, beef scrap to 100, alfalfa to 50. Scratch feed early in the morning and late at night. Green food at noon if not plenty of tender greens on range. Scratch feed same as (E) except change oats to 75 whole oats, or you can give a variety by feeding a mixture as follows:—100 cracked corn, 100 wheat, 50 barley, 50 whole oats. Be sure they have all the scratch food they will eat at night. See that they have full crops. Special attention to your green food during August, September and October. If your range does not supply enough green food you must have a crop ready. Rye and rape make the best quick growing green crop.

FIRST FOUR WEEKS IN HOUSE (G)

Birds should be housed as soon as they show signs of laying, or at all events before cold weather.

Should be fed the same as on the range, except when they start to lay, add to your growing feed, 10 pounds gluten, 5 pounds oil meal, 3 pounds salt. Should have green food at noon. If the birds don't go to roost voluntarily, put them on at night. Watch the litter and keep it clean and dry. Keep down the lice with lice powder and spray.

AFTER FOUR WEEKS IN HOUSE (H) (Heavy Laying Rations)

Change growing feed to following dry mash, or better to moist mash—100 bran, 200 middlings, 200 cornmeal, 200 ground oats, 160 beef scrap, 50 alfalfa meal, 20 gluten, 10 oil meal, 6 salt, 20 charcoal. As they increase in laying, add to last 6 items, so that when birds are in full laying, the formula will be 100 bran, 200 middlings, 200 cornmeal, 200 ground oats, 200 beef scraps, 75 alfalfa meal, 50 gluten, 25 oil meal, 8 table salt, 25 charcoal. This is best fed as warm moist mash, and best moistened with milk, sweet or sour. A few boiled beets, turnips, or potatoes can be added occasionally. Feed in V shaped troughs at 11 A. M. and give them what they will clean up in thirty minutes. There should be enough troughs so that all can feed at once. Scratch feed fed at 7 A. M. and 4 P. M. being sure that they have full crop at night. Seven quarts per 100 hens each feed is about what they will require. In cold weather, can substitute whole corn for cracked corn for night food.

FOR BREEDERS

The mash should not be so heavy. Formula should be 200 bran, 200 middlings, 200 cornmeal, 150 ground oats, 150 beef scraps, 75 alfalfa meal, 25 oil meal, 8 salt, 25 charcoal. Feed moist what they will eat in twenty minutes. We leave out gluten for breeders. Make the mash a little more bulky, and do not feed as much. Breeders should have more yard room and more exercise, which means that more of the food given them is necessary to maintain their body and less goes into the eggs. The feeding of laying stock must be varied more or less to fit conditions. It is useless to feed a heavy ration if the birds will not produce the eggs, (in fact, it is a detriment), and on the other hand, you cannot expect to produce the eggs if you do not give them food to produce the eggs with. If you give them an excess of food and they do not use it in producing eggs, it will go into fat and into waste, and when it goes into excessive fats the hens become lazy and the egg production is again reduced.

Year old hens in summer and early fall can have a mash somewhat lighter than growing pullets and can be made lighter by adding bran. In summer the hens can have less corn and more wheat in their scratch food. If beets or cabbage are used for green food for layers, feed what they will clean up in three or four hours. If sprouted oats, see special article on same.

CAUTION

It is more or less dangerous to feed an unselected flock representing the average from your breeding, no matter how good the breeding, a ration such as used in egg laying contest (on picked birds.) If such formulae are tried, watch results carefully. Often it is advisable to separate what you consider your heavy laying type and feed them differently.

WATER

Of course fresh water, shells, charcoal, and grit should always be before the birds in large quantities.

RENEWAL OF FLOCK

A flock of heavy layers are not economical after the second year. It is economy to renew half your flock each year, because if you wait until all your flock is two years old, it requires twice the brooder and range capacity. The sale of your two year olds as meat should pay expense of raising pullets to maturity. Be sure your new stock was not bred from forced layers. Only by attention to this detail can you repeat the first two years' success.

MOTTO

Your motto should be, see how quick you can get a bag of grain into your chicks, rather than how long you can make it last. This does not mean to give them all they will consume but means to consider the birds as economical machines and feed them all you can convert into useful products. If you feed them only enough to maintain life they will have no material from which to produce profitable products.

CONDENSED MASH FORMULA

CONDENSED MASH FORMULA

	Chick Mash	Growing Feed	Broilers	Range	1st 4 wks. in house	Early Laying	Heavy Laying	Breeders
BRAN	100	100	50	100	100	100	100	200
MIDDLINGS	150	150	150	150	150	200	200	200
CORN MEAL	100	100	125	125	125	200	200	200
GROUND OATS	50	100	100	100	100	200	200	150
BEEF SCRAP*	50	75	100	100	100	160	200	150
GLUTEN					10	20	50	
ALFALFA MEAL	20	25	25	50	50	50	75	75
BONE MEAL		25		25	25			
CHARCOAL		10	10	10	10	20	25	25
SALT		1		1	4	6	8	8
OIL MEAL					5	10	25	25

*OR FISH MEAL

POUNDS OF DIGESTIBLE NUTRIENTS PER 100 LBS.

	Protein (P)	Carbohydrates (C)	Fats (F)
WHEAT	8.8	67.5	1.5
CORN	7.8	66.8	4.3
OATS	10.7	50.3	3.8
BRAN	11.9	42.0	2.5
MIDDLINGS	16.9	53.6	4.1
OIL MEAL	30.	32.	6.9
CORN MEAL	6.7	64.3	3.5
GLUTEN	21.3	52.8	2.9
G. OATS	10.7	50.3	3.8
MEAT SCRAP	66.	0.	13.0
ALFALFA MEAL	11.	40.	1.

SPROUTED OATS

Use only good seed oats. Sulphured oats will not sprout properly. Soak oats over night in a tub of warm water. Next morning spread oats to depth of one inch in sprouting trays. Trays should be arranged in racks, tiers six to eight inches above each other. Sixty-five to seventy-five degrees of heat is **necessary** for proper sprouting in ten days. Seventy to eighty degrees is better. Oats kept in low temperature are apt to require two weeks to sprout and to become slimy.

Trays may be any convenient size, three inches deep, covered on bottom with one-eighth inch mesh galvanized iron wire screen. Trays should be thoroughly washed and disinfected (Use Zinoleum) after each using. Oats should be wet once or twice per day to keep them moist. Luke warm water better.

Sprouted oats should be fed when about two inches high. It is the sprouts, more than the greens, that aid birds' digestion. One to two square inches is a day's feed per hen. Fifty-six pounds of oats will cover 4800 square inch of trays.

On the basis of one square foot per one hundred birds per day, six pounds of oats will be required per bird per year, making cost about nine cents per bird per year.

Plant each day the same amount you feed each day. Increase your plantings in anticipation of greater future needs, at least ten days ahead.

Slimy oats are very injurious to birds. There is no excuse for having them.

DON'TS THAT PREVENT BOWEL TROUBLE

Don't leave any unconsumed food about to get sour, for sour food together with sour land is the cause of half of the bowel trouble.

Don't chill the young chicks as this is the cause of much of the other half of bowel trouble.

Don't neglect to clean out the droppings from pens every day.

Don't let chicks out on land on which are stagnant pools of water.

Don't put baby chicks on same land two years in succession.

Don't overfeed and don't crowd.

CHAPTER SIX
CROPS ON THE CHICKEN FARM

While there are, of course, many farms devoted entirely to chicken raising, it is no doubt true that more than ninety per cent of the chickens raised in this country are raised on farms devoted largely to other purposes. In most instances they are simply a side issue in a general farming undertaking. We refer in another chapter to such chicken raising, as constituting a "Back Yard Poultry" proposition. On such farms crops have been the first consideration. Chickens have usually been added at first by the farmer's wife to constitute a source of income for her own use, but in most cases they have been found so profitable, that further additions have been made to the flock until the farm has developed into more or less of a true chicken farm. In some instances the results from chickens have been so good that crops and other matters have become secondary in importance, except so far as they have a bearing on the chicken proposition. On most of these "Back Yard Poultry Farms", crops, or some other line, have been developed to the extent of making the farm self supporting, or nearly so. The addition of poultry has not increased the overhead expenses so that the gross profit from the poultry is practically all net profit.

It is not likely that we shall offer anything in this chapter of value to the farm that has the crop question already worked out; the suggestions we shall make are more intended for the novice, to indicate to him some of the crops that must be raised, if the farm is to be successful to the greatest degree.

MODERN METHODS

The production of poultry and eggs is nothing more or less in the final analysis than a manufacturing business, in which the machinery is represented by the hens, and the raw materials by the food given to the hens. The egg is the principal product; meat, feathers, manure and that which may be produced by manure, the principal by-products.

The wonderful success of modern business may, we believe, be properly attributed to its utilization of products which at one time were termed waste. The more successful manufacturing concerns today derive much, if not all of their profits, from by-products which were once thrown away. Competition has made it impossible for careless or indifferent manufacturers to prosper.

The chicken business, except in a few instances, has not yet reached the point where modern methods are applied as in other manufacturing lines. When we get this manufacturing idea more firmly established in our minds, and when we become seriously ready to adopt modern manufacturing methods, then, and not until then, will we be able to get the greatest returns from the chicken business. We cannot overlook the valuable by-products of the chicken business and make a real success.

THE CASE IN POINT

The case in point in this chapter is the application of the by-product of hens to the crops on the farm, and vice-versa, the application of the crop producing possibilities of the farm to the needs of the hens. There are many crops that can be raised on any farm that are of great value to the hens, either for food or for shade. Hens, each year, produce a very considerable amount of manure which should be turned into products that they themselves may need, or it should be turned into money with which to supply these needs. If no crops are grown on the farm, this manure is wasted: it is even worse than wasted, for the land gradually gets polluted by too much manure, so that good stock cannot be raised on it.

The records of one government experiment station shows that each hen produces an average of forty pounds of clear hen manure per year during the hours that it is at roost. This is in addition to the still larger quantity of manure that becomes mixed with the litter in the pens and which is known as strawy manure,(than which there is no better coarse manure for general farm use) and that manure which is deposited in the yard outside. All of this manure can be utilized. The best results from the strawy manure, of course, are obtained by ploughing it into the land, but the hens grind up the straw so finely by their continual scratching that it is fairly satisfactory to apply as a top dressing, especially on grass lands. The manure in the yards can be taken advantage of by having the yards so arranged that they may be periodically cropped. The clear manure from the roosts may be used in any way desired.

VALUE OF HEN MANURE

The real value of hen manure as a fertilizer is not generally understood. Most farmers realize in a sort of indefinite way that it brings unusually quick and satisfactory results but they are not well informed as to its value in relation to other manures, nor as to why it does produce these quick results. In bulk, of course, on the ordinary farm, it is much less than the horse or cow manure; as a matter of fact, to put it plainly, what quantity is usually available does not look as if it would amount to much and it is used in a hit-or-miss way. Few people take as much pains to get the value out of their hen dressing, as they do out of the other kinds of dressing that they have. On a farm, however, where there are any number of hens of consequence, particular attention should be paid to this product. On the basis of seventy-five pounds of manure per bird per year, at a commercial value of only $13.15 per ton, each bird produces over fifty cents worth of manure per year. It deposits on the dropping boards alone over twenty-five cents worth of manure per year. Surely this is quite an extra profit that many of us throw away. If one can turn this waste product into value, he gets a profit that is ordinarily hidden, a profit that is easily overlooked without realizing it.

Every farmer should know the comparative value of different manures. We give below table showing the comparative value of different manures, taking a chemical fertilizer for comparison. A good chemical fertilizer contains forty pounds of nitrogen, one hundred and seventy-five pounds of phosphorus, and forty pounds of potassium. This very rarely sells for less than $30.00 per ton, and nearly always much higher. In the calculations that follow, we have assumed that one pound of nitrogen has a money value equivalent to three pounds of phosphorus, or three pounds of potassium,

POUNDS PER TON

	Nitrogen	Phosphorus	Potassium	Comparative Value per ton.
Good Commercial	40.	175.	40.	$30.00
Cow	6.8	3.2	8.0	3.52
Horse	11.6	5.6	10.6	4.47
Sheep	17.6	4.6	13.4	6.34
Hog	9.0	3.8	12.0	3.84
Hen	32.6	31.8	17.0	13.15

It will be noted from the above comparison that hen manure in its original state is not a well balanced manure for general use. The ratio of nitrogen to phosphorus and potassium should be reduced. It is the presence of this large amount of nitrogen, of course, that makes hen manure valuable for inducing quick growth, or for giving quick start to such crops as corn. For best results, however, with the ordinary crop, it should be mixed with other ingredients to bring up the proportion of phosphorus and potassium, or the ground should be so manipulated that it will bring about the same results. For instance, we have raised very excellent crops of potatoes, crops of over four hundred bushels to the acre, with clear hen manure as the sole fertilizer, directly applied, but the results were obtained by ploughing into the ground a good, heavy crop of clover. We do not mean, perhaps, to say that the clover actually supplied the lacking chemical elements, but it did put the ground in such shape, by furnishing new humus, that good results were obtained.

To get the greatest benefit from hen manure, it should be mixed with other materials and we suggest the following formula as one that we have found very satisfactory:

   30 pounds Hen manure
   10 pounds Sawdust or loam
   16 pounds Acid phosphate rock
   8 pounds Kainit

The acid phosphate and kainit are straight, rough chemicals, the former supplying phosphorus, of course, and the latter, potash. They are comparatively inexpensive, prices varying from $8.00 to $12.00 per ton. The way to mix these materials is by making a pile spreading a layer of each material, until the full amount of each is in the pile, the pile to be finally shoveled over and thoroughly mixed together. This mixture under ordinary circumstances should be applied at the rate of about two tons per acre.

THE SOIL AND ITS NEEDS

It is not enough that the farmer should be acquainted with the percentages and character of the ingredients in the fertilizer that he is about to use. There is little value in knowing all this, unless he also knows, first—what his soil requires to produce good results, and second,—what chemicals are best suited to produce proper growth in the crops he is about to raise.

If we should devote this whole book to the subject, we could not give a complete story of the soil and its needs. We shall attempt to touch upon only a few of the more important matters, so that the novice will have a basis upon which to continue a study into the detail of the subject.

The soil should be regarded as a reservoir containing the elements on which the crops are to feed. The upper soil, which we plough and harrow, is what we ordinarily think of as the soil, but the term soil properly applies to the earth to a greater depth than this and this soil in scientific parlance is divided into two parts, namely, the top soil and the sub-soil. The sub-soil, although given much less thought and attention by the farmer, plays a part with many crops equally as important as the top soil, more especially with those crops that have long roots that can penetrate it.

Soil consists principally of disintegrated rock and of humus. Humus is a word commonly spoken and perhaps not as well understood. In order that it may be clear what humus is, we will state that it is that part of the soil that is formed by the decomposition of vegetables and animal matter.

That part of the soil formed by disintegrated rock may not at the moment be in proper shape for plant food. If, however, the land is properly tilled so that air and water may get to it, the elements in the rock will oxidize and combine with the carbonic acid of the air to make a compound which is available as plant food. In case of many of the elements that the plants need, such as iron, sulphur, silicon, etc. the disintegration and chemical change goes on sufficiently rapidly so that the farmer may never need supply any.

The humus of the soil, however, must be replaced from time to time. It serves a double purpose; it makes the soil physically light and opens it up to receive air and water; and it further supplies a large number of bacteria which are continually active in producing plant food. The sources of new humus are manure, cover crops, roots and sod, ploughed into the ground.

Common sense will point out that anything growing on a piece of land must necessarily receive its nourishment from the land to a large extent. The only other source of nourishment is the air, and the elements of the air are in most cases made available as plant food only through the agency of the soil. It follows that the raising and taking off of any crop from the soil must impoverish the soil.

The soil may today contain all the elements, or plant food needed for the particular crop to be planted. It may still further have had these elements made into available food to be supplied to the plants by proper tillage and proper irrigation. At the same time, it goes without saying, that while many of the less important elements may continue to be available as fast as succeeding crops require, there will be some elements taken out to such a degree, that they will have to be promptly replaced to retain the growing capacity of the soil. This replacement will have to be made either by

       1.—Direct Replacement
       2.—Rotation of Crops
       3.—Combination of the two.

The four elements that are most required for plant food, and consequently the most rapidly used, are **nitrogen, phosphorus, potassium and calcium.** Some crops contain more of one element and other crops more of another. In order to properly select a fertilizer for a particular crop, one should have at least a rough idea of the quantity of each element that that crop removes from the soil. In order to give assistance in this respect, we append herewith a table, showing the general requirements of the more common crops.

(Pounds taken out per acre, average crop)

	Nitrogen	Phosphorus	Potassium
Corn	77.2	22.8	64.3
Wheat	31.6	9.7	17.9
Oats	31.9	11.6	36.1
Barley	45.5	15.3	51.4

So far as possible one should select for a particular crop a fertilizer that is rich in the elements most present in that crop. By this means the soil is supplied with that which is to be taken away from it, so that there will be no loss in its producing capacity.

Another method of making a direct replacement of the elements taken from the soil is to grow crops of clover, Canada peas, buckwheat, or some similar rough crop to be ploughed into the soil upon maturity, thereby replacing vegetable humus. Animal humus, is, of course, replaced by the ploughing in of animal manures.

RETENTION OF SOIL STRENGTH BY ROTATION OF CROPS

While it is impossible by rotation of crops to do away with the necessity of replacing the elements that the crops remove from the soil, a good system of rotation enables one to handle his land so that no one element becomes exhausted. In fact one crop often puts into the soil just what another crop requires. It is also equally impossible to get the best results by use of farm manure, or cover crop alone. Some chemicals must always be used to keep up the strength of the soil. A careful study of soil and crops should be made so that the proper chemicals are supplied. In order that our readers may have information regarding the common sources of the elements most required, we give the following:

COMMON SOURCES OF NITROGEN:
    Nitrate of Soda—(Contains about 320 pounds nitrogen to the ton)
    Tankage—(Contains about 120 pounds nitrogen and about 200 pounds Phosphoric acid to the ton)
    Cottonseed Meal—(Contains about 140 pounds nitrogen to the ton)

COMMON SOURCES OF PHOSPHORUS:
    Treated Rock—(Contains from 200 to 350 pounds phosphoric acid (available) to the ton)
    Treated Bone—(Contains from 400 to 600 pounds phosphoric acid to the ton)

COMMON SOURCES OF POTASSIUM
    Kainit—(Contains about 250 pounds potash to the ton)
    Muriate of Potash—(Contains about 1,000 pounds potash to the ton)
    Sulphate of Potash—(Contains about 1,000 pounds potash to the ton.)

COMMON SOURCES OF CALCIUM

The common source of calcium is lime (oxide of calcium) and is found in limestone, marble, chalk, bones, shells, etc.

Lime is removed from the soil to some extent by the crops, but is also removed by drainage. Lime is necessary to the land, however, for other purposes than to supply that element to the crops.

Lime assists in making nitrogen available; it opens up heavy soils and allows air and water to penetrate it; it also binds light soil. Most important of all, it corrects acidity in the soil.

To determine if lime is needed, purchase from the drug store a few cents worth of blue litmus paper and insert it into a handful of the soil you suspect; if it turns red in a few minutes, you will know the soil needs liming.

Lime should be applied after ploughing and before harrowing. It would be put too deep into the soil, if applied before ploughing. No rule can be given as to the quantity to apply, as different conditions require from 200 to 1,000 pounds to the acre and some very acid, clay soils as much as 2,000 pounds to the acre.

Basic slag is also a source of phosphoric acid and contains about three hundred and forty pounds to the ton, but it is not in a readily available form. It also contains considerable lime and is especially useful on sour or clay lands. (See chapter on "Fruit.")

HEN MANURE SUPPLIES ELEMENTS

On the chicken farm, where large quantities of hens are kept, there will be comparatively small need of buying outside elements. The analysis given above shows that it is much richer in the three important elements than any of the other farm manures, and the quantity available is usually more than sufficient for the amount of land owned. To get the best results, however, from this manure, it is desirable to add the cheap, rough chemicals suggested in the formula given, in order to have a better balanced fertilizer.

CHICKEN CROPS

As we said in the beginning, it is not our purpose to give advice regarding crops to the farm whose main business is to raise crops, but we will make suggestions regarding those crops that are best adapted to chicken farms and further suggestions as to the best manner for handling these crops in relation to the needs of the chickens. With a proper rotation of crops and hens, a small amount of land will successfully raise a considerable quantity of both. There is a sort of continuous process, a repetition of the same thing—over and over again, and is illustrated in the old saying:—

> No grass, no cattle;
> No cattle, no manure;
> No manure, no grass.

In what follows, we will assume that conditions are such that only a limited amount of land is available, and we will endeavor to show the point where there is danger of the land being overcrowded. It is quite possible to so use a large piece of land that in a few years it will be less capable of supporting a flock of hens than would have been half that amount of land, it if had been properly rotated. If one should continuously raise a flock of birds on the same piece of land, the ground would soon become so polluted that it would be useless to put young chicks on it. In such cases, of course, it does not matter how large an acreage of land is owned, if only this one small part of the land is used.

FRUIT AND CHICKENS

In the next chapter we discuss the combination of fruit and poultry, showing how the two go hand in hand. We consider this the nearest to the ideal combination. A few crops, of course, may be, and probably would be, raised on a fruit farm, or some fruit can be well raised on a farm not especially adapted for, or devoted to fruit.

GRASS AND CHICKENS

Next to the fruit and chicken combination, undoubtedly comes grass and chickens. The second year of grass is best. Chicks that range on grass land, however, should be provided with shade near at hand. Corn makes excellent shade. To get the best results from the chickens and to keep the land in most suitable condition for the chickens, and at the same time to raise crops that will prove of real money value, a four year rotation is recommended as follows:

> First Year— GRASS
> Second Year— GRASS and CHICKS
> Third Year— CORN
> Fourth Year— OATS.

A CONCRETE EXAMPLE

In order to work out this rotation in an intelligent manner, we will consider a concrete example:

Let us take as a unit, a farm on which it is intended to keep one thousand layers. We will figure the number of chicks that it is necessary to raise each year to keep up the flock on this assumed unit of one thousand layers, and then we will figure the areas necessary to devote to crops, so that the chicks may have new, sweet land each year for their range.

The areas that we will suggest are the minimum that we think can be used and kept in condition, so that the farm will be in as good condition after any one year, as it was the year before. If the land is heavy, larger areas should be used, and in any case, larger areas are desirable, if available.

Of course, if it is intended to keep two thousand layers, twice the area would be required.

EIGHT EGGS — ONE LAYER

Our grandfathers had a rule, regarding the raising of chicks, which used to run about as follows:—

> Eight eggs make FOUR Chicks:
> Four Chicks make TWO Pullets:
> Two Pullets make ONE layer.

It may seem prohibitive to the beginner that it takes eight eggs, or four day-old chicks, to make one layer, but it is not prohibitive because everyone is up against the same proposition.

ONE-HALF NEW FLOCK EACH YEAR

Hens pass their greatest usefulness as money producers after the second year. It follows that a flock kept up to the best producing capacity must be entirely renewed in two years. In other words, one half the flock must be renewed each year. So, on our one thousand layer farm, it is necessary to raise five hundred new pullets each year. On the basis given above, it will take two thousand day-old chicks, to raise these five hundred selected pullets. After the broilers are sold off, there will be approximately seven hundred and fifty chicks to put on range.

ONE ACRE — ONE THOUSAND CHICKS

When new, sweet land is used each year, one acre will properly range one thousand young chicks to maturity, so our seven hundred and fifty chicks would require three-quarters of an acre. Some poultry men, desiring to sell their goods at any cost, regardless of whether the conditions are right where their goods are going, maintain that one thousand chicks can be raised year after year on one acre, and they lead many people to try to do it. After one has a little experience, however, in ranging chicks on land that has not been properly rotated, he gets the idea that ten acres are none too much for one thousand chicks. It is a fact, however, that two acres, properly rotated, will raise our seven hundred and fifty chicks year after year with excellent results. As we have said above, we are stating the minimum amount of land that we think can be used; to be perfectly frank, we would prefer to have three acres instead of two. We are convinced that very few realize the importance of fresh land for raising chicks and keeping hens. We have traced out case after case where poor results have been obtained with chicks, with the resulting conviction that probably fifty per cent of the losses with young chicks is due to infected soil.

DIVISION OF LAND

It is improbable that we can describe here just the shape of the piece of land the reader may have available for chicken purposes. As a basis for our argument, however, we will assume that the land is in a strip five hundred feet long by one hundred seventy-five feet wide. We will divide this land into five equal parts. It could be divided into parcels each one hundred feet by one hundred and seventy-five feet, and this division would probably require less fencing and probably less labor to look after. We prefer, however, that the divisions should be made five hundred feet long by thirty-five feet wide. With this division of the land, the colony houses would be located on a strip of grass thirty-five feet wide. The adjoining strip would be planted to corn, which would furnish shade for the birds. The houses would be in one long row and equally distant from the shade. They should be so placed that the morning sun will strike them.

The following diagram illustrates the method suggested for handling the two acres. We will assume that the land is all grass to start with and you will remember that we have divided the two acres into five parts, each thirty-five by five hundred feet. We have numbered the divisions 1 to 5 as follows:

1.	
2.	CORN
3.	GRASS AND CHICKS
4.	
5.	

You will note that in our plans for the first year, we have planted No. 2 division with corn, and have put the chicks on No. 3. When we come to the second year, we will plant No. 2 with oats, sowing grass seed at the same time, plant No. 3 to corn, and put the chicks on No. 4. In order to make the matter clear and simple, we will illustrate the rotation by the following additional diagram:

	Plot No. 1	Plot No. 2	Plot No. 3	Plot No. 4	Plot No. 5
1st Year		Corn	Grass / Chix		
2nd Year		Oats	Corn	Grass / Chix	
3rd Year		Grass	Oats	Corn	Grass / Chix
4th Year	Corn	Grass / Chix	Grass	Oats	Corn
5th Year	Oats	Corn	Grass / Chix	Grass	Oats
6th Year	Grass	Oats	Corn	Grass / Chix	Grass
7th Year	Grass	Grass	Oats	Corn	Grass / Chix

Etc. Year After Year.

CHICKS LIKE CORN

It may be found advisable to use temporary fences to keep the birds off those divisions of the land where an attempt is being made to grow crops for feed or for the market. In such cases, however, a three foot fence will be found sufficient, as the birds will work from preference near the corn or the house, and will have little disposition to go farther away. Once the corn, oats or other crop is well started, they will do very little injury, but it is not safe to allow them to range on grass land where one wishes to raise hay, until after the hay is harvested.

SUPPLEMENTARY DETAILS

It is impossible to take up the subject of this chapter in as complete and detailed a way as we would like. Nor is it easy to handle the matter in such a brief way in an especially logical manner. We have, however, attempted to take up those points that, we believe, are of the most importance to the chicken farmer, and will supplement them below with brief remarks as to the best way to handle several of the crops that are ordinarily raised.

A few beets and cabbage should be raised each year for winter feed. Cabbage should be fed first, as they do not keep as well as the beets. If the eggs produced are to be used for hatching purposes, beets and cabbages should supplement, rather than displace sprouted oats. We believe both forms of green food have special value for breeding purposes.

PLANTING DIRECTIONS

Beets—

Beets require a soil that can be ploughed and mellowed ten inches deep. They should be planted in rows eighteen inches apart. Plant eight to ten pounds of Mangel Wurtzel seed per acre. Use the equivalent of one thousand to twelve hundred pounds of fertilizer per acre. When the plants are well started, thin down, so that they stand not less than ten inches apart. If manure is used, it will be found profitable to add four applications of nitrate of soda, each application at the rate of one hundred and twenty-five pounds to the acre. The seed should be sown one inch deep and as soon as the ground can be worked. For hens, we have found the Golden Tankard the best variety. For winter use, the beets should be stored in pits or in sand in the cellar.

Cabbage—

This vegetable requires deep, rich, moist, but well drained soil. Cabbage for winter use should be planted late. Where seeds are planted, they do not need to be put into the ground before the first of June. If they are started in seed beds, the cabbage can be set out on a piece of land that has already been used during the same season for a crop of peas, that is, they would be set out about July 10. If the planting made is to be permanent, they should be in rows three feet apart, and when well started, should be thinned out so to be one foot apart. An acre planted to cabbage so that the plants are one foot apart one way, and three feet apart the other will yield fourteen thousand five hundred and twenty heads per acre. As in the case of beets, applications of nitrate of soda are beneficial, but three applications will be found to be enough. The first at transplanting time, the others one and two months later respectively. New land is especially well adapted to growing cabbage.

If one has not a suitable cellar for storing cabbage, where they can be kept at an even temperature, he can adopt an outside storage method that works well. Select a piece of ground well drained, and set the cabbage with roots up, on this ground: then cover them lightly with leaves and as the weather gets colder, continue to add sufficient leaves to prevent freezing. A few boards will shed the water. Another method commonly used, is to dig a hole on a dry knoll, or other dry place, putting the cabbage into it, and covering with boards and dirt. This, however, is nothing more or less than an improvised cellar.

Ground for planting cabbage should be thoroughly stirred with cultivator and hoed every week until the plants cover the ground. If the young plants are troubled with the cabbage fly, sprinkle them with tobacco dust, air-slacked lime, slug-shot, or wood ashes, while the dew is on them. To keep off the cabbage worm, sprinkle with Dalmatian insect powder or slug-shot.

Turnips—
Turnips may be sown at all seasons from April to August in northern climates, although those will be the best which are sown very early in the spring for summer crops, and early in August for a fall and winter crop. A light soil, well manured the previous year, is the best. A few hundred pounds per acre of quick acting fertilizer will have a tendency to make the roots smooth and to make the rapid growth so essential in producing sweet and palatable turnips. Newly turned sod will produce the sweetest and smoothest turnips. Turnips furnish a good variety of green food for hens in the winter. The amount of turnip seed required per acre is approximately two pounds and the seeding should be done in the same way as in the case of beets.

Carrots—
Carrots also make a very palatable food for occasional feeding in the way of variety. They should be sown in rows twelve to eighteen inches apart and three-quarters of an inch deep on rich soil that has not recently been manured. Early carrots should be sown in April, and late carrots May to July. Early varieties require about seventy-five days to mature, late varieties one hundred to one hundred and twenty days. Carrots are practically free from damage by pests.

Corn—
Corn requires a good soil and a warm situation. Yellow corn should be planted as early in the spring as the ground becomes warm. Sugar corn in the early varieties should be planted about the first of May and if a continuous supply is wanted all summer, plantings should be made about two weeks apart from the first of May to the last of July, first planting early variety, and then a later one. The seed should be planted two inches deep in hills three feet apart. Some authorities say that five kernels in each hill is plenty. We believe that twelve seeds to the hill is better. When the plants get started these should be thinned down to three plants to the hill. Late varieties should be covered a little deeper than the early varieties. It requires sixty to ninety days for corn to mature.

The amount of corn seed usually required per acre is eight to twelve quarts of field corn and eight to ten quarts of sweet corn.

In the rotation recommended in the chicken farm, the corn should follow the second year grass crop. It will be planted, of course, primarily for shade for the chicks, but nevertheless will supply considerable grain for later food, as the chicks do not damage the crop to any great extent. Some years when our suggested rotation is followed, there will be two strips of corn, only one of which will have the chicks in it. It is sometimes advisable to use a very slight amount of commercial fertilizer to give the corn an early start, but this depends largely on the condition of the soil.

Oats—
Oats will follow the corn in our rotation. The grass that follows the oats can be sown at the same time. We consider oats one of the most profitable crops that the chicken farm can raise, because every part of it is available for the chickens' use. The straw makes the very best of litter and absorbent, while the oats make excellent food. We consider

oats the best growing food for young chicks, but of course, they must be furnished to the very young chicks finely broken up, in such form as oatmeal. Chickens cannot handle whole oats sucessfully until at least half grown.

There are many different ideas of the proper sowing of oats and grass. It goes without saying that the land should be well ploughed and thoroughly harrowed. Oats take away a good deal from the soil in the matter of chemical elements and the ground should consequently receive a liberal dressing. Owing to the fact that the oats will lodge if the ground is too rich with the ordinary manure or fertilizer, it is advisable to add to the usual application of manure about three hundred pounds to the acre of a specially prepared commercial fertilizer, made up largely of nitrate of soda. There are two alternatives, if this extra fertilizer is not added, namely:

To dress the land so that the following hay crop will be satisfactory, and run the chances of the oats being spoiled by lodging, or

To dress the land lightly, to make sure of a good crop of oats, but at the same time limiting the production of hay that may be expected later.

In our practice we have been liberal in our application of dressing and supplemented it with the special fertilizer. The results have been excellent. Some people make sure that their oats do not lodge by giving a very light seeding of one to two bushels to the acre. We consider this a penny-wise policy. The best formula that we have used for combined oats and grass seeding, is as follows:

> 10 quarts Timothy
> 4 pounds Alsike Clover
> 4 pounds Red Clover
> 3 bushels Oats

Other formulas that we have tried with satisfactory results are as follows:

CLARK FORMULA

> 3 to 3½ bushels Oats
> 14 quarts Timothy
> 14 quarts Red Clover

MAINE EXPERIMENT FORMULA

> 5 pecks Oats
> 10 pounds Timothy
> 7 pounds Red Top
> 6 pounds Alsike

another formula is:

> 1 peck Timothy
> 3 pounds Alsike
> 3 pounds Red Clover

The formulas above represent the quantities per acre.

If the ground to be seeded should need liming, the lime should be applied just after ploughing, usually at the rate of about one ton to the acre. The first year's grass will be largely clover and being unused by the chickens, will be made into hay. The second year the grass will be largely Timothy. If the chicks can be kept off that until late, the Timothy may be made into hay, before they have a chance to damage it. The best time to apply clear hen dressing to grass lands, is as a top dressing between the time the oats are harvested and the first grass crop; *i.e.*, during the winter.

We believe the above represent the main crops that the ordinary chicken farmer will attempt to raise for the actual use of his chickens. But chicken farmers should not forget the vegetable garden, for people must be fed as well as hens. If the farm happens to be on a good automobile route, he can derive much profit from disposal of his surplus garden vegetables to motorists. We will give a few suggestions regarding the planting of the ordinary vegetables.

ASPARAGUS

Asparagus will grow in any soil, but sandy soil is best. The seed should be planted one inch deep and one inch apart in rows eighteen inches apart. The plants should be later thinned so that they will be twelve inches apart in the rows. Plant as early as the ground is fit. The following spring transplant, so that the plants are from two to three feet on centers. Grow for two years in permanent beds before cutting; set plants deep. In the fall, cut off the tops and burn them. Hen manure may be applied in the early spring, and it is best mixed with Kainit. After the cutting season, an application of nitrate of soda can be given. Poultry will keep away all pests if allowed to run through the beds in summer. If rust appears spraying can be resorted to. Common salt is often used in sandy soil, but its value is doubtful.

BEANS

Any soil that is not too rich in nitrogen is adapted to beans. If the beans are planted in rows the seeds should be laid three inches apart, and two or three inches deep. If in hills, plant four feet apart each way, six seeds per hill, thinning to three plants. Plant from May fifteenth to June fifteenth. It requires from sixty to ninety days for maturity. Dwarf beans are the earliest to mature, but the bean considered best by those competent to judge, is the Kentucky Wonder, planted in hills and grown on poles.

GARDEN BEETS

The method of handling table beets differs from the one used for the coarser beets in that they are planted in drills twelve inches apart, the seed to be one inch deep in Spring, and two inches deep in Summer planting. Plant between the first of April and the fifteenth of August. Do the first thinning when leaves are five inches high, using the green leaves and small beets as greens. The first thinning should leave the plants not nearer than three inches, and the second thinning not nearer than six inches to each other. The second thinning will give small beets for eating or pickling. Sixty to ninety days are required for maturity. If arsenates are used to kill pests, do not eat the greens.

CAULIFLOWER

Too difficult for the amateur to grow. Therefore no suggestions are made here.

CHARD

Any soil that is not too wet is suitable. Should be sown one inch deep in rows eighteen inches apart. Thin the plants to three or six inches, eating the thinnings as greens or transplanting them. When the leaves are big enough to eat, the plants can be cut and they will grow again.

CUCUMBERS

Any good soil is alright for cucumbers, but a so-called quick soil is preferable. They should be sown as early as the first of June and may be sown as late as July fifteenth. Plant the hills about four feet apart and sow ten seeds to the hill. Remove all but four plants. The seeds should be sown about one inch deep. There are many pests that are injurious to cucumbers, but if the vines are watched carefully, and sprinkled with some good insecticide, there is little difficulty in getting plants beyond the stage where they may be hurt. Sixty to ninety days are required for maturity.

LETTUCE

Lettuce requires a light, warm and quick soil. It may be sown under glass in March, one-quarter inch deep, or out doors one half inch deep as soon as the ground is fit. It is exceedingly difficult to raise good heads of lettuce from sowings maturing after July fifteenth. Transplanting should be done when the leaves are four inches long, and the plants should be set out nine to twelve inches apart in rows one foot apart. A little nitrate of soda in small but often repeated applications is beneficial. Pests are not troublesome. It requires ninety days for maturity.

ONIONS

Onions might also be included as a food for hens but should be used sparingly. They can be used to good advantage once a month.

Plant in rows a foot or less apart for hand culture, or thirty inches apart for horse culture. Sandy soil is not well adapted to onions. The soil should be medium loam, rich and deep. Sow one-half inch to one inch deep and thin to three inches when the plants are three inches high. Onions should be constantly cultivated, especially after each rain, and hand weeding is necessary at the time of thinning. Onions require much dressing. Hen manure may be used up to as much as ten tons per acre. Occasional dressings of nitrate of soda or liquid manure are desirable. They may be planted under glass in March and transplanted when of sufficient size. Plant it in April, as soon as the ground is fit, if seed is planted in the garden.

PARSNIPS

Parsnips require the whole season to mature and should be planted as soon as the ground is fit. They may be left in the ground through the winter. They require a rich, deep soil, but should not be planted in ground that has been recently manured. After they cover the ground well they do not need further care. Plant one-half inch deep and in rows eighteen to twenty-four inches apart.

PEAS

Plant as soon as the ground is fit, about April first. Plant early peas one-half inch apart and two inches deep. But later in season plant in trenches five inches deep filling in dirt around them as they grow. Plant in double rows six inches apart with eighteen inches or more between the double rows. Thin to two or three inches and pinch off ends if growing too rank. Dwarf varieties are best for ordinary planting as they do not need any support when planted in double rows. Soil should not have an over supply of nitrogen as that produces rank growth. Peas are best planted on land that was highly dressed the previous season.

POTATOES

Cut the seed potatoes into about four pieces each with one or more eyes on each piece, and plant about a foot apart and in rows twenty-four inches apart for hand cultivation. Plant four to six inches deep. Grass land freshly turned is well adapted to potatoes. Plant April fifteenth to May fifteenth. Fertilizer should be applied broadcast before planting or it can be put in the rows under the seed if first covered with a layer of earth. For large plantings, pests, etc. see treatise on potatoes and Government bulletins.

RADISH

Soil should be light and quick. Sow one-half inch deep in drills six inches apart and thin to two inches. Plant every ten days from April first to June fifteenth. Can be sown again in late August.

SPINACH

Spinach wants a rich, light, quick soil, rich in nitrogen; frequent applications of nitrate of soda is recommended. Sow one inch deep in rows twelve inches apart. Thin to four inches. Begin to sow as soon as ground is fit, or from about April tenth to September fifteenth. New Zealand Spinach will bear all summer and is best planted in hills four feet each way. The seed should be soaked over night as it is slow in starting. It should also be sown one inch deep, about a dozen seeds to the hill. Eight hills will supply the average family all summer.

SQUASH

Soil should be quick, well drained and warm. For summer squash the dwarf Crookneck is best. Sow in hills one inch deep and four to five feet apart. Thin to four plants to the hill. The young plants should be well sprinkled with Bug Death or other bug destroyer. In winter varieties pinch back the vines if they get too long and allow not over three fruit to the vine. Ground should not receive any fertilizer after planting.

CHAPTER SEVEN
DUAL FARMING

FRUIT AND CHICKENS THE IDEAL COMBINATION

To have "too many irons in the fire" is usually bad policy. This is as true in farming as in other lines. As a rule those who make a specialty of some one thing are the most successful. There are, however, exceptions to this rule, especially in those instances where one line of work may prove a help and complement to another. Many farms are unprofitable simply because too many branches are attempted with the capital available; or because there are more things to look after than the ability of the farmer can master. On the other hand many farms would be brought to a paying basis if some other, well selected branch were added.

TREES AND CHICKENS HELP EACH OTHER

The combination of chicken farming and fruit farming is an apt illustration of this latter class. The combination is ideal because each needs the assistance of the other. Each needs what the other will supply. In the case of chickens an outlet is needed for the dressing produced, shade is required, insect animal food is necessary and a factor of economy in labor is needed. In the chicken business expert help is required and this help must be carried through the year: one cannot afford to let a good man go, so it is most profitable to have some line of work to consume the unused labor hours. Fruit farming takes care of all these needs.

The fruit farm needs dressing and can employ chickens most successfully to destroy insects and for minute and continuous cultivation of the ground about the trees. The fruit farm furthermore needs the help of some other branch to help defray the expense of development.

START DEPENDS ON CAPITAL

It is fair to assume that many people who contemplate starting in the chicken business or the fruit business have not sufficient capital to make a real start in both. In such cases the real start should be made in one and a partial or gradual start in the other. A few hens may be added to the fruit farm and increased as circumstances permit, or a few trees may be planted each year on the chicken farm. We believe this course should be pursued for best final results. In this chapter, however, we will take it for granted that the farmer has sufficient capital to make a full start in both branches at the same time.

FACTORS TO CONSIDER

In selecting a farm for this combination of chickens and fruit there are several factors that should be considered.

DRAINAGE

First, air and soil drainage are most important. For peaches, pears, plums and the more perishable fruits a hill top with a gentle slope to the southeast is best. In cold northern climates where the main fruit is such apples as Baldwins, a slope to the northeast is generally considered preferable. The most important consideration, however, it seems to us, is to have the whole farm well elevated above the surrounding country, an elevation of one hundred feet is ideal. Under these conditions the frosty air will drain down into the valley and in winter, when a hard freeze comes after a thaw, the trees are less likely to be damaged and the newly grown buds will be subject to less injury from late spring frosts. In northern climates north slopes are more desirable for the apple because the sap will not make so many false starts in the warm days. Of course slopes that are too steep are not desirable on account of washing of the soil and on account of the extra labor entailed.

SOIL

A deep, medium loam is the best soil; it may be cultivated easily and will quickly absorb any surface water. Soils of a clayey nature are very illy adapted to either fruit or poultry. A very sandy soil, while excellent for poultry, is not suited to fruit. Any well drained soil is suitable for poultry if there is enough of it; on a fruit farm there always is enough and it is in just the right condition for the best poultry results.

ACCESSIBILITY

The second important consideration in selection of the farm is its accessibility. The ideal location is on or near a trolley line that has express and freight service. If this is not possible care should be taken not to get too far from the shipping depot or from the sources of grain or other materials needed on the farm. Ready access to the railroad means arrival of your products to the consumer or consignee in better condition and a very considerble saving in cost of marketing, besides a very important saving in time during a period when time means most. It is also good judgment, if a farm has to be bought, to buy it in a section near a good market for the products it is intended to raise. Furthermore, it is easier to keep good help on a farm near a town or city where there is opportunity for a little amusement in the off hours.

ARRANGEMENT OF LAND

The combination fruit and poultry farm should be divided into at least two large ranges. There should be a high fence forming the division and the poultry houses should be placed along this fence so that either range may be used at will. Care should be taken that the division is made so that it will not interfere with a proper location of the houses. A division so that the higher land of the farm forms one range and the lower land the other range would be the most natural and the most advantageous as regards both the poultry and the fruit business. The houses can thus be placed so that higher land on one side will protect them from the prevailing winds. We do not mean by this to place the houses in a low spot, but to locate them just under the highest land.

With this arrangement the lower range will contain the heavier and wetter soil, which will be used for the smaller fruit such as raspberries, blackberries, currants, etc., or can be used for hay or such vegetables as asparagus. The higher range will contain the lighter and dryer soil and should be planted with the tree fruits such as apples, plums, pears, etc.

In other chapters we have given details of poultry management so only such suggestions are necessary here as apply to the use of the above ranges and to the breeds adapted to the fruit farm.

COVER CROPS

To properly carry fruit trees through the winter a cover crop of clover, rye, or something similar is necessary as described later. About April first the poultry should be turned into the high range, where the fruit trees are, to clean up this cover crop. The birds are allowed to remain on this range until about August first. This is the period when the insects are the most troublesome and when the shade of the trees is most appreciated by the birds. About August first, or at about the time when peaches begin to ripen, the chickens should be transferred to the lower range. There will very likely be more or less fruit trees in the lower range, but the bulk of the peach or other fruit crop will be in the higher range and cut off from the birds so that the latter will not be in the way during the harvesting. The crops of small fruit in the lower range will have been marketed by this time.

BREED TO CHOOSE

The selection of the proper bird for the fruit farm may very well be left to one's individual taste. We would suggest that we have chosen the leghorn for our fruit farms because they are great foragers. It is important that the birds cover the whole range as far as possible. There is some difficulty about the land near the houses becoming over fertilized. It is well to arrange the ranges so that a part of the land near the houses may be fenced off temporarily if the birds spend too much time on it. The birds can be encouraged to range more widely by feeding at a distance from the house. Frequently however, it seems necessary to consider this land near the houses as waste land to a large degree, avoiding its use for fruit or other crops and reserving its use solely for the poultry as a general meeting ground for all the birds at sundown before roosting time and in the more or less unfavorable weather when the birds do not wander far. The amount of land thus wasted is of small consequence, amounting usually to not more than fifteen or twenty feet. The soil on this patch can occasionally be spaded up and carried off to some other part of the farm as fertilizer, new soil being carted in front of the houses to replace it.

Colony houses for the young growing stock should be placed just beyond the limits where the older birds go.

POINTS ON FRUIT CULTURE

Opinions are by no means united on the best manner of caring for an orchard but as the methods that we have practised on the Fay Mountain Farm, our Westboro Leghorn Farm, have been productive of good results, we will describe the most important points of that method here for the assistance of our readers.

CULTIVATION

In the first place, of course, the land should be gotten into good shape. By this we mean that all large stones should be removed, places that are uneven should be corrected so there may be no places for water to stand, quitch grass should be killed out and so far as possible all weeds destroyed. The soil should be brought into the usual condition for any crop.

The small fruit should be set out in long, straight rows, the rows being sufficiently far apart so a pair of horses can be used between them for plowing and cultivating; possibly at some future time it might be desired to use a tractor.

ARRANGEMENT OF TREES

In the orchard the best method is to plant apple trees thirty-six feet apart each way; then plant between them trees, so-called fillers, that will come into bearing state several years before the apple trees require all the space. Peach, plum and pear trees make good

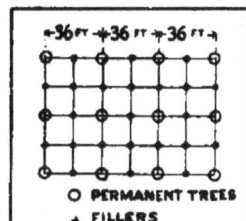

fillers. The final planting by this system will consist of trees eighteen feet apart each way and an acre will contain about thirty-five apple trees and about one hundred other trees. We wish to lay stress on having the rows very straight each way of the orchard; much trouble and annoyance will be saved when the trees get grown.

Some orchardists recommend forty feet as the best spacing for the apple trees, thus making the plantings, including fillers twenty feet apart. This arrangement reduces the number of trees to the acre to one hundred and eight. We have never tried this spacing so can not say whether it would be more economical or not.

DRAW THE LINES STRAIGHT

After the ground has been carefully plowed and harrowed and the stones removed the field should be carefully laid out, placing a stake where each tree is to go. A few hours thus used to see that the lines are straight will save days of labor later in cultivating and harvesting the crops. The next step is to dig all the holes except those on the outside rows. Then set the trees, lining them by the outside stakes. Be sure to thoroughly tamp the earth about the roots of each tree. Finally dig the holes for the trees in the outside rows and plant in the same manner. After planting spread a large shovelful of hen manure around each tree, being careful that it does not touch the tree. If you prefer, one pound of the following mixture may be used to each tree:

    100 pounds Nitrate of Soda
    300 pounds Muriate of Potash
    800 pounds Basic Slag.

In order that fruit trees may get the right start they should be set out, before the ground settles in the spring, while the soil is still muddy. In New England this means by May 1st.

HEN MANURE FOR FRUIT TREES

Our experience has shown that hen dressing will produce more growth per unit than any of the fertilizers or dressings commonly used. The danger in its use in the orchard is that it may produce too much growth; that is that the trees will grow too much to wood at the sacrifice of fruit. This matter requires no watching for the first few years, because growth of the wood is what is desired and nothing will produce it better than hen dressing. When the time approaches, however, when it is thought fruit should be forthcoming and it appears that the tree is putting its energy into wood instead of fruit, all trouble may be overcome by application of Kainit, which is sort of raw potash and which will offset the effects of too much nitrogen. The Kainit should be applied to the ground in a circle about the tree at the point where the ends of the roots may be expected to reach. In an ordinary case of overgrowth of wood about forty pounds of Kainit is a proper application.

Another way, and perhaps a better way under many circumstances, to overcome excess of nitrogen in the soil and to prevent too much growth of wood, is to crop the soil for a year. Grass is probably the best crop for this purpose.

Fifteen years after planting it is probable that the apple trees will need all the room, so that the small plum trees, etc. should be removed. Without doubt at that time these smaller trees will have outlived their usefulness.

The more the orchard is cultivated the better for both the trees and the poultry. Cultivation means more water for the roots, less weeds and fewer pests, while for the chickens it means cleaner soil and more food. Cultivate once each week from the time of spring plowing up to August first.

About this time the poultry is transferred to the lower range and the orchard is sown to the cover crop. Many things are used for cover crops. A good formula is one bushel of buckwheat and one bushel of rye to the acre.

Where buckwheat and rye are used, and where there is evidence of over fertilization in the orchard, the buckwheat crop can be cut and cured about October first, and used as bedding in the laying house. This will still leave the rye for cover. The rye lives over the winter, and gets its growth the next spring after which it is plowed in for humus.

The cover crop is a very important item in proper handling of an orchard. It stops the growth of the trees and allows the new growth to harden so that it can withstand the difficulties of the winter season; it helps to properly ripen the fruit and it holds the plant food for another year. It takes up the surplus richness of the soil and cleanses the soil. It holds the snow, prevents formation of ice and keeps the ground from getting too warm during winter thaws; and not the least of its advantages is the humus it adds to the soil when plowed in.

PRUNING

Pruning should be done in the fall or winter at such times as are most convenient. We have always been in favor of pruning when the sap is not in the tree. There is an old saying, however, that the time to prune is whenever the saw and the knife are sharp, which means at any time when a spare hour for the purpose is available.

In case of raspberries or blackberries only three of the new canes should be left each year, and these should be cut back at least one half. Fruit trees, like apple and peach trees, should be cut back when set out to a whip 18 inches long; after that only three to five branches should be allowed to start from the main trunk. Keep the trees pruned so they will grow into the shape of a huge cup, somewhat hollow in the center, but not so open as to allow the tree to sunburn. The tree, however, needs light and air and attention should be paid to removing all broken or cracked limbs, all bad forks, all diseased branches and practically all branches that grow toward the center or that cross each other or grow downward. Where a cut of any size is made the wound should be covered with good lead paint. Watch particularly for all so-called suckers about the base of the tree for these cause an absolute waste of energy.

KEEP MICE AWAY

In climates where there is snow, wire netting should be placed around each tree trunk to prevent mice from feeding on the bark. This wire should be sufficiently loose so that the tree will not touch it.

SPRAYING

There are three important sprayings that must be made for clean trees and clean fruit.

1.— In the winter or early spring before the sap starts, with one gallon commercial lime-sulphur to nine gallons water.

2.— When the apple tree buds show green at the tips with the same mixture to which has been added one-half pound of Black Leaf Forty to each fifty gallons of mixture. This last is for the Aphis.

3.— Just as the petals drop from the apple blossoms, using four pounds arsenate of lead, and one-half pound of Black Leaf Forty and one-half gallon lime-sulphur to each fifty gallons of water.

All the above sprayings are used on apple trees but only the first is used on peach trees. The peach trees should have for a second spray, at the time the husks drop from the little peach, the self boiled lime-sulphur spray (fifteen pounds lime — fifteen pounds sulphur— fifty gallons water.) This last spray will control the brown or ripe rot. If three pounds of arsenate of lead are added to the above spray this formula will serve as an all round spray and will give protection from insects as well as fungus.

CROPS

A fair crop may be expected from the raspberries and such fruit in the second year, from the peaches in the third year. Apples require from six to eight years before the first crop. It is wisest to remove blossoms from apple trees if they appear before the fifth year.

MARKETING

The growing of the fruit, or eggs, or broilers is only half the game, and often the smaller half. Much of your labor is wasted if your products are sold without forethought and judgment. If the cost of a bushel of peaches is forty cents and you sell at that price, you have simply made both ends meet and you have nothing for the bank. If you sell for fifty cents, you have something, but not all you should have, to put away. If you use a little thought in your selling and get sixty cents per bushel, you have made one hundred percent more profit than you would have made without that thought.

You should study the selling end. Do not let the commission agent or the purchaser make all your prices for you. See what the outside market offers. We make the following suggestions from our experience.

A clean, handy package for your eggs or fruit adds five to ten per cent or more to their value.

Sort into firsts and seconds and have the firsts, real firsts. Sell the top grade at top prices, or a little higher.

Sell to retail stores all that you cannot sell direct to the consumer. Cultivate the automobile trade.

Remember that it takes time to sell to the retail trade and that you should have extra for that time. Remember that this trade is willing to pay the farmer for that time. Remember that this trade is willing to pay the farmer even more than he will pay the store for fresh, first grade products, direct from the farm.

WHERE THE HENS COME IN

The hens help the crops almost as much as the crops help the hens.

They eat millions of bugs; it does not take any imagination to see the difference between the orchard with, and the one without, poultry.

The hens do the work of the smoothing harrow, saving many hours of actual labor.

A thousand hens add dressing equivalent to twenty tons of chemical fertilizer each year.

There are some disadvantages in having poultry on a fruit farm, such as the hens roosting in young trees and breaking the limbs; and there is the danger of overfertilizing, but these can be easily avoided by a little ingenuity. The combination of hens and fruit seems to us to be ideal.

CHAPTER EIGHT

DISPOSAL OF POULTRY PRODUCTS

"A large part of the profits in poultry keeping depends on the marketing of the products. The poultryman must be a good salesman as well as a good raiser of poultry. Either he must have good markets to begin with or know how to make them. High-priced trade is not found readymade" nor can it be "made to order" by others and transferred. The poultryman must make it himself. To build up a high-class trade requires time, skill, tact and high-grade products that somebody wants and that most people will not take the trouble to supply. Such a trade is worth working for. The easiest money to be made in the poultry business or in any other business, is the margin of profit received for extra quality put up in an attractive package, delivered at the right time to the right market."

Prof. James E. Rice.

In spite of the fact that during the spring months, when the American hen is the usiest, the price that is paid in the general market for eggs approaches dangerously near the production cost, there is always an opportunity to get a substantial additional and remunerative price for a good product.

It is probably true that over ninety percent of the hens in this country are kept in small flocks on general farms. Statistics show that the actual average number of hens kept in a flock is fifty-three. The production of eggs in a flock of this size, or smaller, is of such insignificant proportion that the farmer rarely considers it worth while to try to seek a special market for it. The method of marketing usually consists in taking the eggs in small quantities to the country grocery store, there to be exchanged for flour, groceries, etc.

The country groceryman disposes of the accumulation locally so long as the local price is satisfactory. At other times, or whenever he has a surplus, he ships them to the city market. By the time the eggs reach the city market, they are several days old and can in no way be classified truthfully as "strictly fresh" eggs. They are, however, classified as fresh eggs of a certain grade, but as evidence that there is some suspicion as to their real quality, the Chamber of Commerce ruling allows that if they are eighty percent fresh, they shall pass unchallenged. Such eggs in no sense can be considered satisfactory to the fastidious housewife.

AN EXCELLENT OPPORTUNITY

The fact that such a large percentage of the eggs produced are marketed in this way, offers an excellent opportunity for the particular poultryman to build up a special fancy trade for his products. It is only necessary to see to it that all eggs are shipped when freshly gathered, in good, clean condition and in clean packages.

The quality of the bulk of the eggs that reach the market is a disgrace to the industry. If it were really difficult to produce eggs of high quality some excuse might be found. As a matter of fact, however, the production of good eggs is extremely simple and the only explanation for present methods of marketing is indifference. There is no line of agriculture in which high quality is so easily secured. When hens are kept in clean pens, fed on wholesome foods, they will deliver eggs regarding the quality of which there can be no criticism. Poor quality is entirely due, under these conditions, to human carelessness or indifference. No particular skill is required on the part of the poultryman to produce a good product.

STUDY THE MARKET

It, of course, requires some slight knowledge of market conditions to suit the best trade of that market as regards size and color of eggs, and the most satisfactory manner of packing, but this knowledge is nothing but what can be acquired, if a little interest is taken in the

ONE ROAD TO POULTRY SUCCESS

matter. If the poultryman can once gain a reputation of delivering to his customers eggs of unquestionably high quality, his trade will soon overlook any little idiosyncrasies he may have in such matters as packing.

Few people realize how rapidly eggs deteriorate to the extent that they can be no longer classed as strictly fresh. The chief causes of deterioration are lack of cleanliness, exposure to high temperature after being laid, infrequent gathering, (many times not gathered until a broody hen has set on them for several hours,) storing in places where they are subject to odors or dampness, and improper feeding of materials that transmit their peculiar taste to the egg. There are none of these causes but what can be avoided by the careful poultryman.

RULES TO FOLLOW

The rules to follow to supply choice eggs are few and simple, and every beginner should become familiar with the requirements of good quality and the methods to adopt to meet the requirements. We will offer the following few suggestions:

Use Good Food—In the first place the hens should be fed only good, clean, wholesome food of known quality, so that the taste of the egg may be normal. Table waste is very frequently used for small flocks, and danger lies in this practice. Where it is used it should be fed fresh from the table and any waste of particularly strong flavor should be discarded. It does not pay to try very radical experiments in feeding. Often times things cut green from the field will give a peculiar flavor to the egg. We have known it to happen in the case of green oats.

Gather Frequently—The eggs should be gathered frequently, not less than twice a day under any conditions, and in hot weather or freezing weather, much more frequently, even as often as every hour if necessary. Eggs that are found in the litter should never be shipped as a first class product, because one has no knowledge of how long they have been laid. One of the great troubles with the ordinary farmer's eggs is that they are gathered from the haymow, under the doorstep, or the raspberry bush, wherever and whenever they are found, and all packed in together.

Demand Cleanliness—Eggs should never be shipped to particular customers unless they are absolutely clean. It rarely pays to wash eggs. Washing takes the bloom from them. Furthermore there should never be any need of washing them. If the nests are kept clean, it is very seldom that the eggs will get soiled. One pretending to classify his eggs as high grade can well afford to use what few soiled eggs there are at home. Of course, occasionally an egg may have a slight spot or discoloration; this is best removed by a damp cloth.

Grade to Uniform Size—Only those eggs should be packed together which are uniform in size and color. If the hens are laying eggs which are varying greatly in both respects, attention should be given to the selection of breeders, so that more uniformity will be gained. The box or case in which the eggs are packed should be clean and fresh, and it will be found profitable to have a box bearing the distinctive mark of the producer. It seems to us that the method frequently used of stamping on the egg the date on which it is laid is a questionable practice, as many times eggs are not directly consumed.

Store Carefully—Even good sized flocks do not always lay eggs in sufficient numbers so that they may be shipped on the day they are laid, in which case all eggs as soon as gathered should be stored in some place of medium, even temperature, (preferably forty-five to fifty degrees,) and should be left undisturbed. They are best kept in baskets and the baskets should always be clean and sweet.

Ship Promptly—Lastly, do not fail to ship the eggs as soon as you have sufficient quantity to make up the case or other unit of shipment.

STERILE EGGS

There is one special grade of eggs that offers extra profit to the poultryman who will take pains to produce them. These are known as sterile eggs, and are produced by those pens in which no males are kept. The outlet for these is principally in hospitals, although there is other trade that demands them. The advantages of this class of eggs is that they keep much better than those that are fertilized. In the fertile egg, development of the embryo has already begun when the egg is laid, and growth is resumed promptly at favorable temperature. Development of the embryo may begin at a temperature of ninety degrees; such temperature is common in poultry houses in summer, and eggs are often warmed to ninety degrees by exposure to the sun on the way to market. Strictly sterile eggs may be kept at ordinary temperature for weeks, and even months, without developing a strong odor.

A GOOD MARKET

The premium that is offered for strictly fancy eggs should offer much encouragement to the poultryman to seek that market. And it should hold out inducement to the beginner to keep a flock of sufficient size, so that he may ship reasonable quantities of eggs with little delay and so that he can afford to have the equipment necessary to produce a first class product. Even if a beginner should start with a small flock, he should look upon the arrangement as a temporary one. When properly handled no other branch of farming pays so good a return on the investment as poultry keeping, so not only poultrymen, old and new, but farmers as well, should generally increase their flocks to really commercial proportions. With flocks of two hundred or more, labor saving equipment may be provided and proper methods employed; and, during the greater part of the year, eggs may be shipped in regular thirty dozen cases sufficiently frequently to make it possible to dispose of them in the most exacting markets. The production of two hundred hens will pay a very high rate of interest on the investment necessary to put oneself in position to accomplish these results.

TABLE POULTRY

In a general way the remarks regarding the proper methods of marketing eggs apply to table poultry. A very large percentage of the poultry that goes into our market, goes as live poultry, or dressed in such an inferior manner as to make one wonder why the poultryman bothered to dress it at all. It is unfortunate that such poultry should have the influence on the market for good poultry that it does. But of course the price of poultry, like the price of everything else, is governed by the demand and supply, and the price of the best grades has to follow, to some extent at least, the variations in price of the poorer grades. Nevertheless the poultryman who takes pains in the growing, fattening, dressing and marketing of his birds will always be able to get a price considerably in advance of his more careless and indifferent neighbors, and a good profit for his work.

NOT SO SIMPLE

The production of high grade table poultry is not quite so simple a matter as the production of high grade eggs. It is not, however, a difficult problem to master. The wide awake producer will study his market, to know what class of poultry is required for the different seasons of the year. He will follow the quotations to know which size of chickens will bring him in the most profit. He will become familiar with the requirements of a fastidious trade, as to the condition of the birds that will best meet the consumer's tastes; whether they should be picked by the dry method or the scalded method; whether they should be drawn or undrawn, etc.— and he will attempt to meet these requirements to the finest detail.

MILK FED CHICKS

It is safe to say that no one wants poor, scrawny chickens. They must be well meated, filled out in every section. The flavor and quality of the meat must also be up to a certain standard. The condition of the chickens in this respect depends almost entirely on how and what they are fed. It will probably be found that milk-fed chickens are the best. Milk-fed chickens are those that have been fattened on a special ration of ground grains, mixed with buttermilk, or skim milk. They may either be crate-fed or pen-fed. Crate-feeding means that the chickens are confined in small coops and fed certain definite quantities at stated intervals during the day. We question if there is any great advantage in this system. Pen-fed chicks usually turn out a good quality if they are closely confined. Milk-feeding with close confinement results in rapid gains, better flavor and a marked softening of the flesh. We will not discuss this matter of feeding here as it is taken up more fully in the chapter on "Feeding."

Neither does the particular customer wish chickens delivered to him spotted with pin feathers, or with the flesh torn by careless picking. Such poultry is far from appetizing and does not encourage the use of poultry for table use. Particular care should be also taken that the legs and comb and beak are thoroughly cleaned.

CATER TO GOOD TRADE

No housewife is so indifferent that she does not appreciate the delivery to her of any article in a nice, clean package. The delivery of specially fattened fowls, well drawn, and prepared for the oven, wrapped in waxed paper and neatly tied up, cannot fail to make an impression on her, and create a desire in her to have more of the same quality at another time. The cost of such care in catering to one's trade is trifling and the returns on the investment will be large.

POULTRY CLASSIFICATIONS

There are various sizes of poultry that are in popular demand. The market classifications of these in the ordinary market are:

- Squab Broilers
- Broilers
- Fryers
- Roasting Chickens
- Soft Roasters
- Capons
- Fowls
- Stags

SQUAB BROILERS weigh from three-fourths to one pound each and are served whole as a substitute for quail or pigeon. They are also in demand as a substitute for small game birds.

REGULAR BROILERS weigh from one to two pounds when dressed. The season for both squab and regular broilers, when good prices can be obtained, if from December to May. There are, of course, large quantities of broilers offered in the market after May. These are the surplus cockerels produced by egg farms.

FRYERS weigh from two and a half to three and a half pounds, and are mostly supplied by the farms of the country. They are the chickens that have run on open range all summer, and are as a rule hard fleshed and thin. They are generally of poor quality, although there is no need of them being so.

ROASTING CHICKENS include all chickens weighing three pounds or over. Like the fryers the great bulk of the supply comes from the farms, and the quality is similar. Each year, however, sees a greater number of roasters coming from well managed poultry plants, and in this stock the quality is very much improved.

SOFT ROASTERS. This term originated on the South Shore of Massachusetts where it is applied to a special grade of fowl that are so bred and handled that they are brought to a large size while still soft meated and tender. The chicks are hatched in the fall, raised in brooder houses until they no longer need heat, and then transferred to Colony houses. They are hopper fed and are offered no inducement to take exercise. The originators of this type of poultry have been very successful in the production of them.

CAPONS are castrated male birds. The operation is performed usually when the cockerels weigh about two pounds. The effect of the operation is to make the birds quiet and contented, so that they grow to a larger size and remain soft meated indefinitely. The production of capons is limited and the opportunity for successful business in them is great.

FOWLS are the hens that have passed their usefulness as layers. While easily disposed of at fair prices, they are never of as good quality as roaster chickens, even after special effort has been made to improve the quality.

STAGS are old roosters and cockerels that have passed into the breeding stage. It is practically impossible to improve their quality. They bring the lowest prices of any form of poultry.

There are so many complete and valuable books on the best ways for killing poultry and preparing for market that we will not go into the matter in this chapter. One of the best books that we know of on picking poultry is one entitled, "Martin's Way of Picking Poultry" by H. L. Martin, Marblehead, Mass.

CHAPTER NINE
GENERAL TOPICS

The more information one has about his business the better understanding he will have of it, and the more likely he is to make a success of it. This chapter is only a series of brief remarks on a variety of subjects, all of which are more or less related to the poultry business. The different suggestions are just so many shots fired at random in the hope that some of them, at least, will hit the mark and be helpful.

WHAT AN EGG CONTAINS

A whole new-laid egg contains nearly thirteen percent protein, a little more than 10½ percent of fat, nearly 66 percent water, more than 10½ percent ash and a trace of carbohydrate.

The yolk, which comprises 32.75 percent of the whole egg, is 17½ percent protein, nearly 33 percent fat and a little more than 48½ percent water. The albumen, or white, which comprises a little over 57 percent of the whole egg, is 12.3 percent protein, 0.6 percent fat, 0.2 percent ash and somewhat over 86 percent water.

The shell and its membranes, which comprise about 10 percent of the entire egg, contain about 93 percent calcium carbonate, 1 percent magnesium carbonate, 3 percent calcium and magnesium phosphate, and 3 percent organic matter.

WASTEFUL MARKETING

As an incentive to the novice to start in business with a firm resolve to use intelligent and business-like care in the marketing of his products, with consequent greater profit to himself, we quote the following from an article written by a prominent food chemist whose words have unusual weight.

"The egg is one of the most valuable of human foods, yet in the face of an increasing population and lowered food production, and in spite of loud complaints about the high cost of living, we Americans waste $50,000,000 worth of eggs every year, while of those that do reach the consumer in an edible condition, only a very small percentage can be eaten with anything like satisfaction. In the latter case it is estimated that the loss from deterioration amounts to another fifty million. There is a tremendous additional loss from breakage. This, according to the estimates of the Department of Agriculture, amounts to one dozen in every thirty, making a total in the case of New York City alone of 116,000,000 eggs annually."

It seems hardly credible that poultrymen can carry their carelessness so far as to be responsible for such stupendous economic waste and the knowledge of these facts should compel each one of us to do what he can to better these conditions. The facts, also, offer ample explanation as to why the intelligent and prudent housewife is willing to pay a good premium for eggs that she can rely upon as being one hundred percent fresh and one hundred percent in otherwise good condition.

Save the feathers if you do killing on the farm.

BROODER SUGGESTIONS

As a great percentage of all chicks under modern poultry culture is raised in artificial brooders, a few suggestions regarding things to do, or not to do, in brooding may be helpful. In view of the fact that we consider the coal-heated colony brooder much the most desirable brooder to use, all suggestions will be made from that standpoint.

The first three weeks are considered the most important in a chick's life: it is surely worth every effort to get them safely through this period. It is not a very difficult matter, if one is not afraid of work and is willing to devote himself conscientiously to little details. Failure is usually due to carelessness or indifference, or perhaps to thoughtlessness.

It is first important, of course, to have a good brooder. We shall assume that the reader is the owner of one of the many good types that it is possible to get. As good results in brooding depend to a considerable extent on proper heat, one of the chief things to look for in a brooder is a good regulator. It is our rule at least, to start with a temperature of 100 degrees under the hover during the first few days. Then to gradually reduce the heat so that at the end of three weeks it is about 85 degrees. The regulator should be set to keep the heat at these marks. It is possible that these temperatures are too high to maintain under the hovers of the old fashioned lamp heated brooders, which are so limited for room, but there is little danger of overheating a colony brooder, because the chicks have a wide range of temperature to select from and have ample room to get away from the point of greatest heat, if they so desire.

IT IS IMPORTANT that there should always be some part of the brooding space that is sufficiently hot to satisfy the chicks greatest demand for heat. The same temperature is not always comfortable for every chick any more than it is for every human being. The chick may become chilled at a heat that would appear uncomfortably warm to another. Chicks must, above all things, be prevented from becoming chilled, as chilling is the cause of a considerable part of chick troubles, bringing on, among other things, a bowel trouble. which while perhaps not as disastrous as white diarrhoea, is still responsible for great losses. It is safe to say if the heat goes wrong everything is useless.

THE USEFULNESS OF A GOOD REGULATOR lies in its keeping the same proper heat during both the day and the night. A poor regulator is a poor servant, because it usually fails in its service when your back is turned. It need hardly be said that low temperature does as much harm in the night as at any time. Unless it is quick acting and positive and well constructed, the regulator will not overcome the natural variations of the fire during the several hours in the night when the fire does not have human attention.

It is equally important to select a brooder with ample coal capacity for the same reason, that is, unless the stove will hold sufficient coal for a good night's service, one can hardly expect a proper heat to be maintained. No regulator will take the place of coal.

GOOD AIR AND VENTILATION is absolutely essential for the chicks' welfare. It is not sufficient just to have a good brooder. Do not impose an unreasonable task upon it. Care must be taken that the brooder is installed in a proper building of sufficient size and with adequate provision for circulation of air. We believe in muslin front buildings for young chicks as much as for fowl, but modified to suit the needs of the younger birds. We have taken up this matter so thoroughly in the Chapter on Poultry House Construction that nothing more needs to be said about it here.

CLEANLINESS IS IMPERATIVE in handling young chickens, as in all things. If chicks do not have clean quarters with clean surroundings they cannot thrive. Cleanliness is as necessary for the ground outside of the building as for the floor inside. There should never be a bad smell in a brooder house.

The litter in the pen and under the hover should be raked over every day and all accumulated droppings removed, and all waste food removed. If waste food is allowed to lie around on the floor, it will become sour and cause disease. The litter should be changed frequently, how frequently depending upon how much the birds are confined to it. Food hoppers, food boards, and drinking vessels should be frequently cleaned and scalded. The whole pen should be entirely cleaned out twice a week at least and, in some instances, as often as every day. It is only by keeping everlastingly after things that one can hope to have success.

CHICKS SHOULD ALSO BE PROVIDED with plenty of room for exercise. The colony brooder system offers more opportunity for exercise than any other. While there is, of course, a larger number of chicks together than in any other system, each chick has a greater latitude to range over. Not only should the inside brooder accommodations be ample, but the yard accommodations as well. It does not do to confine chicks too closely, they simply become stunted. There is the same relative need for exercise for chicks as there is for children.

Of course the matter of feeding the chicks is of the very greatest importance but this is taken up in such detail in another chapter that we will omit all mention of it here.

Before you make a permanent practice of killing and dressing your birds for market, try shipping them alive. Under certain circumstances the latter method proves more profitable.

WHITE DIARRHOEA

White diarrhoea is probably the one chicken disease that poultry men talk most about. It is not nearly as prevalent, however, as many poultry men, or at least as many inexperienced poultry men, suppose. There is a tendency to call any bowel trouble by this name. We shall not attempt to give any scientific discussion of this disease, but merely touch upon the methods by which it is transmitted and upon the possibilities of checking it.

MATURE FEMALE FOWLS are the original source of white diarrhoea. The disease is conceived in the ovary of the mother hen. The ovary develops the yolk of the egg. Danger of transmission to young chicks comes when the eggs are affected by the spores of the parasite. It follows, where the opportunity for infection exists, many chicks have the disease when hatched.

WHITE DIARRHOEA, however, may be transmitted in other ways. Infection may take place from chick to chick during the first three or four days of their lives. This is the critical period. The droppings of the affected chicks contaminate the soil, food and water. When the droppings become dry, they crumble and are blown about by the wind. Infected dust, containing the disease spores, is deposited in the food or water or green grass out doors where the chicks are kept, and outbreaks of the disease follow. The spores are not destroyed by water. Protection of food and water against contamination is of the very greatest importance.

INFECTION spreads among the older fowls through contact with infected droppings. If the soil in the yards is contaminated, standing pools of water become the greatest source of danger.

THE FIRST WEEK OF INFECTION is the most fatal. If the chicks live through this period deaths are not so frequent. Chicks with strong vitality may recover, but do not as a rule make satisfactory growth. If they are females they may harbor the disease and become a source of infection later.

WEAKNESS AND LACK OF VITALITY are the prominent symptoms of white diarrhoea. The chicks become listless, sleepy and sometimes droop their wings. They huddle together under the hover, and lose appetite. In severe cases they stand around with their eyes closed and become indifferent to everything around them, except that many will chirp constantly as if cold or in distress. The discharge from the vent is white or creamy in color, sometimes mixed with brown, frequently stopping up the vent. With practically no exceptions death of chicks from white diarrhoea will occur within a month of being hatched.

A DISEASE SO WEAKENING and so ravaging should be fought with all possible skill and persistence. Sanitation should be the by-word. The best protection against it is to breed only from sound stock.

THE AVERAGE POULTRY KEEPER will find it difficult to tell whether or not his flock is infected with this disease, but he may get positive information on the subject by having his birds examined by the experts at the Agricultural colleges. The only safe course to pursue in breeding is to select for breeders only the healthiest and most vigorous hens. It is good practice to disinfect all eggs used for hatching by immersing them in a solution of one gill of Zenoleum to eight quarts of water.

SOUR MILK seems to be the only easy agent for controlling white diarrhoea. It suppresses intestinal putrifaction which the parasite of this disease sets up.

ON ACCOUNT OF THE RAVAGES of white diarrhoea, especially during the first few days of the lives of baby chicks, it is important to begin feeding sour milk early where the presence of this disease is suspected. Moreover, sour milk should be constantly kept before the chicks until the dangerous period of the disease is passed.

Have work laid out in advance so that the rainy days will be the busiest days.

FOOD ECONOMY

There are many ways of practicing economy in the poultry business and every possibility of practicing economy should be investigated. There is probably as great an opportunity for saving in the feed bill as in any other single item. Effort should be made to have money on hand to purchase quantities of the different grains at the times of year when they are cheapest, which is usually in the fall and early winter. Ground feed should be mixed as soon as bought and stored in boxes or barrels. There is considerable waste in dipping feed out of bags. Hoppers should be so constructed that the fowl will not waste the grain. A lip on the front of the hopper will prevent the hen from dragging the mash out with her beak.

Intelligence should also be used in the kind of food that is given to the hens, effort being made to supply that which best meets the hen's requirements, and not to supply that which the hen will not eat.

There is such a thing, however, as false economy in feeding. It is not economy to curtail in the quantity of food given the hens, if they can assimilate more and turn it into salable products for you. Our rule is to see how quickly we can get a bag of grain into the birds, rather than to see how long we can make a bag of grain last. The more grain that you can get the birds to eat without getting fat, the more eggs you will get, which of course is good economy, if you are after eggs. Also, the quicker you get a certain amount of grain into the hens that you are fattening the less grain it takes to do the fattening. The idea is that it takes a certain amount of food to replenish the waste of the body, and if you give the hens only this amount, there is nothing left with which to produce eggs or fat. The birds need to be watched carefully, however, because it is quite possible to overfeed. Do not get too much corn in your ration if the birds show signs of getting fat. We refer you to the chapter on "Feeding" for examples of balanced rations.

Sell the birds that are past their usefulness today, and not next week.

If you desire to build up a good automobile trade in eggs, see that the eggs are located handily so that you can get a package ready without running all over the farm. If you have other products to sell, have them displayed so that the egg customers will see them.

FORCED LAYERS DO NOT MAKE GOOD BREEDERS

The egg farmer, whose sole purpose is to produce market eggs in the largest quantities, can ignore the laws for handling poultry that the breeder finds it imperative to follow. Outdoor runs, non-stimulating diet, adequate floor space, special methods for inducing exercise, are of no particular concern to the egg producer. Heavy feeding, limited exercise, close confinement, and other things that lead to heavy egg production are extremely injurious to breeding stock. Breeding stock should come to the breeding season in the best possible condition, with ample reserve force, and the ability to transmit great vigor to the chicks. Birds that have been forced for egg production through the winter period are considerably weakened, their vitality is low, and they are consequently unfit to produce good hatching eggs. The policy of using the same flock for heavy egg production and for breeding cannot be too strongly condemned. If an early tendency to lay is noted among those birds intended for breeding, it should be discouraged.

The necessity for taking seriously the above suggestions has been proven many times not only by experiments under Government supervision, but by practice of experienced poultrymen. The facts also expose the fallacy of trying to breed heavy layers from heavy layers, provided the breeders used have accomplished big productions as a result of forcing methods. As a rule, two hundred egg hens are produced by hens that laid hardly more than half of this number of eggs, rather than by hens that had themselves reached the two hundred mark. It will be well for the novice to consider this matter carefully, in order that he may make no mistake in purchasing his stock. The best stock will be secured from a parent stock that is kept solely for breeders in the best of breeding conditions.

Attractive methods of putting up products pay big returns.

A small sign set up in a sightly place and arranged so that it may be changed day by day as the products to be sold change, will bring in a good many extra dollars.

DISPOSAL OF DECAYED ANIMAL MATTER

Many destructive diseases among poultry can be attributed to the chicks or fowl eating decayed animal matter. Limberneck, common in the South, is directly traceable to this. Many poultrymen are careless about the proper disposal of dead animals. The utmost care should be exercised to promptly remove any chickens, hens, or other animals that die, so that the chicks or fowl may by no chance get at them. In want of a better method, they may be disposed of by burial, but unless buried deeply, there is danger of the bodies being scratched out, or of maggots working to the surface. There is only one safe and sure way to dispose of the bodies, and that is to burn them. On some of our plants, we have a concrete pit arranged for this purpose.

INCUBATORS AND INCUBATION

Incubation is a science in itself. In modern poultry keeping it is becoming more and more a business in itself. To properly incubate eggs, so that they will produce chicks of really satisfactory quality, requires so much knowledge of the science and such frequent attention to the eggs and machines during the incubating period, that it is hardly worth while for the ordinary breeder, to say nothing about the breeder of small flocks, to devote any time at all to hatching in these days, when chicks of the very finest quality can be purchased, all hatched, from reliable breeders, or when one can take his own eggs to a mammoth custom hatching plant, where the work will be done with the best of equipment under the most expert care at a very reasonable cost.

In discussing incubators and incubation we are going to assume that most poultrymen, at least that class that have made a success of the business, are ready to concede that it is unprofitable for the small breeder, or any breeder who does not hatch at least twelve hundred eggs at one time, to spend any of his time on incubating. There are poultrymen, of course, who prefer to do their own hatching for the pleasure that there is in it, but this whole book is written from the viewpoint of making profits, so we can take no recognizance of those interested in the business from any other standpoint. So in what we say below, regarding the hatching problem, we take it for granted that the poultryman has in mind a hatching capacity of at least twelve hundred eggs at a setting. The first thing to consider is the incubator.

WHAT IS A GOOD INCUBATOR

The chief requirements of a good incubator are

1.—That it will maintain absolutely even heat.

2.—That it will have adequate provision for ventilation.

3.—That it will not hinder the application of sufficient moisture.

4.—That it will require a small amount of labor for operation.

TWO TYPES

From a practical standpoint there are only two types of machines that we need to consider, the small lamp heated machine and the large, hot water heated, so-called mammoth, incubator. Let us first consider the small machine, to see how it meets the requirements of a good incubator mentioned above.

THE LAMP HEATED MACHINE

Practically all small machines derive their heat from a kerosene lamp. Nearly all are equipped with good working regulators, which will do their work well so long as the lamp behaves well. If, however, the lamp varies in its burning, due to wrong adjustment of the wick, or poor oil, or because of strong drafts of air, or if the lamp goes out, of course there will be a wide variation of temperature, because the regulator is not capable of overcoming the deficiencies of a poor heat supply. As a rule, however, there is not much trouble in maintaining a reasonably regular heat in modern small machines, but it is always something that the poultry men will worry over as liable to cause difficulty at any time. During the third year of our experience in the business, we operated thirty 400-egg machines. We do not recall that we went through a single full hatch without difficulty with the lamp on some machine.

VENTILATION

In the matter of good ventilation the small machine is considerably deficient, especially when compared to the possibilities of the mammoth incubators. In the first place there are always the fumes from the oil lamp to contend with. The air is never pure and free. Furthermore, it is impossible to change the volume of air that the embryo chick may have to draw from, because the eggs are the same distance from the source of heat at all times.

THE DRY SYSTEM

With few exceptions, manufacturers of small incubators recommend that they be operated under what is known as the dry system and no provision is made for moisture. We do not believe that eggs can be properly hatched without moisture, at least we were never able to hatch chicks in sufficient numbers or of sufficiently good quality without it. To overcome the lack of moisture in our small incubators we made a practice of keeping pans of water in the nursery drawers, under the eggs, up to the eighteenth day, but as the chicks dropped into these drawers after hatching, it was necessary to get along without the water after the eighteenth day, which period was the most important time to have it.

WASTEFUL ON LABOR

In the matter of labor, small machines can in no sense be considered economical. It requires practically as much time to care for a small machine of four hundred egg capacity as for a mammoth machine of sixteen thousand egg capacity, outside of turning the eggs. Whether you consider the labor cost on the basis of small capacity or of a large capacity acquired by duplication of small units, the cost per egg for hatching is very much greater in the small machines than in the big machines. Even the cost of turning the eggs is greater in the small machines, as each and every egg has to be rolled or turned by hand, whereas, in mammoth incubators, a whole trayful of eggs is turned by simply placing an extra tray over the eggs and inverting the trays.

MAMMOTH INCUBATORS

Modern mammoth incubators are made up of rows of compartments, one after another, all of the same size, and each holding about one hundred and fifty eggs. The machines may be as long or as short as one wishes between the extremes of twelve hundred egg capacity and twenty thousand egg capacity. It is our opinion that machines on either side of these limits do not operate well. Whatever the length or capacity of the mammoth machine, the heat is supplied from one source, namely, from a small hot water heater at one end of the machine, hot water from this following coils of pipes which run in the top of the machine over the egg compartments, and exposed in the egg compartments.

THE DIFFERENT STYLES

There are various types of mammoth machines with respect to regulation of heat. Some machines have a regulator on each egg compartment and each compartment has to have separate regulation. Other machines do all the regulating at the heater. Still another type is regulated in both ways.

The machine selected for our own use, after careful investigation, and the machine that we have used through many years of experience with the utmost satisfaction, is that type built by the Hall Mammoth Incubator Company of Little Falls, New York. The chief reason why we selected this machine was on account of this very matter of regulation.

The method adopted by that concern makes use of a regulator on the heater, which maintains an average constant temperature for the whole machine. As, however, there is always some variation in the temperatures of different compartments, when eggs of different ages are in the different compartments, some method had to be adopted to permit of separate regulation for each compartment; but it was important that whatever method was adopted should not entail too much labor in its operation.

ADJUSTABLE TRAYS

The Hall people adopted the plan of having trays that could be adjusted at different distances from the pipes. This not only permitted the having of perfect temperature in each compartment, but it allowed the further very decided advantage of being able to gradually move the developing chick farther and farther away from the pipes, allowing it more volume of air, as more air was needed. This system of regulation has never given us the slightest trouble and as can be readily seen, it has a considerable advantage over other systems on account of this matter of air supply.

AIR SUPPLY

With reference to air supply, there are, of course, no gas fumes of foul odors to contend with in Mammoth machines. The cellar may be thoroughly ventilated with no worry about drafts disturbing the heat supply. The machines may be operated so that each compartment has practically as much or as little air as is desired.

MOISTURE

Owing to the fact that Mammoth machines are built with an open bottom, it is a simple matter to supply adequate moisture. This may be done by simply keeping the floor wet under the machines, but this method has the disadvantage that the moisture is uneven, because of evaporation, unless it receives the very closest attention, the floors being at times very wet, and at other times more or less dry. In our own plant we have a shallow sand pit under each machine, in which the sand is thoroughly wet once each day. The sand prevents rapid evaporation.

AN ILLUSTRATION

In the matter of labor, as we have suggested above, there is no comparison between the small machine and the mammoth machine. We will take for illustration a hatching plant of sixteen thousand egg capacity. If it were equipped with largest size of small incubators, the four hundred egg size, there would be forty of these machines, requiring the filling, cleaning, and trimming of forty kerosene lamps each day. It would require at least six hours each day to turn the eggs. With the mammoth machines there is only one small heater to attend and the time required for turning the eggs would not be over two hours.

In other words, on a poultry plant having sixteen thousand egg hatching capacity the small machines would require that there be a man to devote his whole time to nothing but hatching, while the large machine would permit this same man to put in several hours of work each day in some other part of the plant. Furthermore, in one case there is always the risk of accidents to the eggs and the risk of the hatching of inferior chicks, while in the other case the risk is practically nothing.

AN IMPORTANT POINT

A matter to which the purchaser of a large incubator should give careful consideration is the construction of the heater. The success of a machine depends, of course, upon having ample and reliable heat. The heater should be well constructed and particular note should be made that the fire pot is larger at the grate than at the top, so that the fire will not arch.

HINTS ON INCUBATION

Incubators should be located in a cellar, because there is less variation in the room temperature of a cellar than of a building above the surface of the ground. Uneven temperature in the incubator room will make it difficult to operate the incubator at an even temperature. Cellar conditions permit of better regulation of moisture. A cellar should be constructed so there will be sufficient light to work by and ample opportunity for ventilation

---

A large loft over the cellar, and separated from the cellar by a floor or ceiling, is a decided advantage in hot weather, when the sun beats upon the roof. When there is no such loft, the sun often makes it so hot in the cellar that it is impossible to keep the heat in the incubators down, even if the fire in the incubator heater is allowed to go out.

Windows should be so located, or at least so equipped, that the sun will not shine directly on the incubator.

Nothing should be kept in an incubator cellar that will cause disagreeable or musty odors. The incubator cellar is a poor place for vegetables.

None but strictly fresh eggs, carefully selected for good normal shape and color, should be set. We have made a great many experiments with eggs of odd shapes and in various unusual conditions, and in no case have we ever had satisfactory results from them. Poor eggs simply take up room in the incubator that should be occupied by good eggs.

Although we have experimented at different times with hatches in which we turned the eggs only once per day and from which we had good results, we nevertheless feel that it is advisable to turn the eggs twice, the two turnings as nearly twelve hours apart as possible. In handling the eggs, the operator should be careful not to jar them unnecessarily, and his hands should always be clean.

Attend to the incubator heater at regular periods and give it as nearly as possible the same kind of attention each time. Be sure that the heater is kept clean, especially that it is free from clinkers.

We have found it desirable to disinfect the eggs before setting them. We accomplish this by having a pan sufficiently large to take in a whole trayful of eggs. In this pan we mix a solution of Zenoleum in the proportion of one gill of Zenoleum to eight quarts of warm water, and immerse the tray, loaded with eggs, in the solution, allowing it to remain only for an instant.

This same Zenoleum solution is an excellent disinfectant with which to wash the trays and compartments after each hatch. Do not fail to see that everything is thoroughly cleaned about the incubator after each hatch.

HOW TO MAKE A POST MORTEM EXAMINATION
(Taken from Cyphers Company Bulletin, No. 13, 1913)

Even if most of the normal symptoms have been noted in the flock of sick chicks, it is not always safe to conclude that bacillary white diarrhoea has made its appearance, until after the diagnosis has been confirmed by a post-mortem examination. Every regular or professional poultryman should learn to make such examinations, and it will prove to be a good investment to sacrifice a few healthy chicks in order to learn exactly how the internal organs appear when in a normal and healthy condition.

The easiest way to make the autopsy is as follows: Procure a large shingle or board of white pine or other soft wood, into which tacks or push-pins can easily be pushed. Place the chick on the board, breast uppermost, and stretch out the wings and legs, tacking them in this position. Slit the skin covering the breast and abdomen, then peel it back sufficiently to expose the breast and the muscular wall of the abdomen. With a knife or scissors make an incision below each side of the breast bone and remove the entire breast. This exposes the internal organs without disturbing them in any way.

In typical cases of death from bacillary white diarrhoea these conditions will be observed.

The chick is usually emaciated, having only a small amount of muscle on legs, wings and breast.

The crop may be empty, or partially filled with food or slimy liquid. Sometimes the latter is present in considerable quantities.

The lungs are normal. In our investigation we have not observed any of the tubercles that are a prominent symptom of aspergillosis, but other investigators state that occasionally chicks are found which are suffering from both aspergillosis and bacillary white diarrhoea.

It is most important that the condition of the liver be carefully noted. This organ will usually be pale, with streaks and patches of red. These congested areas are not clearly and sharply defined, but are easily observed. In occasional cases the liver is more or less congested throughout. The kidneys and spleen seem to be normal.

The intestines are pale, showing a dirty white color instead of a healthy pinkish white. Usually they are practically empty, though at times a comparatively small amount of gray or brown matter is found.

The ceca are rarely found to contain firm or cheesy matter, but are either empty or partially filled with soft material of a grayish color.

The unabsorbed yolk is usually present, in size ranging from that of a pea to a full-sized yolk. There is a great variation in its consistency. Sometimes it is normal, at others very liquid or watery, and occasionally gelatinous. It usually has a stale odor, but is not putrid in fresh specimens.

To recapitulate, the more prominent indications of bacillary white diarrhoea are: Sleepy, peeping chicks having a glairy, white discharge; many showing "Short backs," a high mortality from the fifth to the twentieth day; livers, pale with red patches and streaks; intestines, pale in color, usually empty; unabsorbed yolks. If all of these conditions are observed, the poultry man will be justified in concluding that bacillary white diarrhoea has made its appearance in his flocks and he should take steps accordingly.

HOW TO CLEAN A POULTRY HOUSE

The following excellent advice is given by Dr. Raymond Pearl, of the University of Maine

"Not every poultry man of experience even, knows how really to clean a poultry house. The first thing to do is to remove all the litter and loose dirt which can be shoveled out. Then give the house-floor, walls and ceiling a thorough sweeping and shovel out the accumulated debris. Then play a garden hose, with the maximum water pressure which can be obtained, upon floor, roosting boards, walls and ceiling, until all the dirt which can be washed down easily is disposed of. Then take a heavy hoe or roost board scraper and proceed to scrape the floor and roosting boards clean of the trampled and caked dressing and dirt. Then shovel out what has been accumulated and get the hose into action once more and wash the whole place down again thoroughly and follow this with another scraping.

Next, with a stiff-bristled broom thoroughly scrub walls, floors, nest boxes, roost boards, etc. Then, after another rinsing down and cleaning out of accumulated dirt, let the house dry out for a day or two. Then make a searching inspection to see if any dirt can be discovered. If so, apply the appropriate treatment as outlined above. If, however, everything appears to be clean, the time has come to make it really and truly clean by disinfecting. To do this it is necessary to spray or thoroughly wash with a scrub brush wet in the solution, used for all parts of the house, with a good disinfectant at least twice, allowing time between for it to dry."

TWO COMMON CHICKEN DISEASES
CATARRH AND ROUP

Catarrh is common to chicks and fowls of all ages. It takes the form of bowel catarrh in little chicks and of head catarrh in older birds. The form that is, however, most common and harmful from an economic standpoint, is the head catarrh that appears in cockerels and pullets that are just maturing.

Head catarrh is usually the result of colds that appear in the flock of young birds in the fall before they have been taken in from the range. Sometimes it is brought on by sudden changes of temperature in the late fall after the birds have been housed, and is especially noticeable in unsanitary houses. However, it results chiefly from allowing chicks to remain on range too late. It is a common occurrence for the ordinary poultry man to put off his work so that the winter quarters are not ready for the new crop of birds as early as they should be, necessitating the leaving of the new birds out in the small houses on range through the usual period of fall rains.

A slight discharge at the nostrils is usually the first symptom of a cold. This discharge will frequently get mixed with dust, so that a crust is formed over the nostrils, causing the birds to breathe with difficulty. The birds will be found going around with their mouths open.

If proper attention is given to these colds, they should not prove at all serious, but if proper attention is not given, they are quite likely to develop into roup, which is a very serious disease.

Roup is easily detected in pronounced instances by a swelling of the eye lids, due to bubbling of air and the secretion of mucous at that point, and by what is known as the "roup breath" which is adequately described only by the word "rotten." When one gets familiar with this odor of roup he will never question the presence of the disease as soon as he enters the pen where it is present.

The transition from a catarrhal cold into roup is gradual and it is impossible to distinguish accurately between cases of just common cold and roup. As roup, however, is something that every poultry man should make the most earnest effort to avoid, he should put forth his efforts to take care of the common colds and thus in all probability make his most efficient effort to keep his flock free from roup. When you detect the smell of roup, "Get Busy."

There are many remedies for roup and cold. Nearly all supply-houses sell a Roup Remedy so-called, which has sufficient curative qualities to be worth a try-out at least. We have found it profitable to swab the throats of the birds with kerosene and to pour kerosene into the nostrils. The application in the same way of a good coal tar disinfectant, like Zenoleum or Napcreol, is beneficial. Arsenite antimony in doses of 1/1000 grain three times a day to a bird of ordinary size is a good remedy for colds. Where the cold is prevalent through a large part of the flock, one should figure out the dose necessary for the flock and mix it with the drinking water.

GOING LIGHT

Another disease which occasionally makes its appearance among the maturing birds in the fall is what is termed Going Light. This is very similar to the tubercular trouble in human beings commonly called consumption. It is doubtless due to lack of constitutional vigor, or perhaps to poor nourishment or exposure.

There are no decided symptoms such as may be observed in most other diseases. In this respect also it is like the consumption of human kind. There is a gradual losing of flesh and thinning of the blood. One may perhaps express the course of the disease by saying that the bird gradually wastes away, until it becomes mere skin and bones and barely able to stand up.

There is no satisfactory treatment for the trouble. It rarely affects more than a few specimens in a flock, so considerable trouble will be saved if the poultry man will just put the few specimens that there are out of their misery.

While there is no cure, fresh air will doubtless prove a good preventive. There is probably no disease with which the beginner feels more helpless than with Going Light.

Keep in mind that in a flock of one thousand layers every egg by which you increase the yearly average lay per bird means $25 more profit for you. An increase of one dozen eggs in the yearly average production means an extra $1.00 per day to you.

HOW TO SET HENS

Provide a nest from twelve to sixteen inches square and from fifteen to eighteen inches high, according to the size of the breed that you are setting. The nest should not be too tightly shut in on top. It should have a slatted door or cover, however, so the hen may be kept in the nest.

Short hay or short straw, or any fine material like them, may be used for making the nest. It should be well pressed down before setting the hen, unless you give the hen a few china eggs for a day or so.

The nests should be located in a room by themselves. In the afternoon let all of the hens that were set on the same date out on the floor at the same time for food and drink, and also for exercise. It is not necessary that they be allowed to remain off the nests more than fifteen or twenty minutes. It will be better if the hens occupy different nests when they return. While the hens are off the nests, the nests should be cleaned and all broken eggs and other refuse removed.

It is best to feed only hard grains to setting hens. Whole corn is one of the best foods because it takes longer to digest. A limited amount of green food and water should be provided. For anything like satisfactory results a careful watch must be kept for lice. The hen should be dusted two or three times during the incubating period, so that the chicks will not be infested when they are hatched.

It is wise to test the eggs on the seventh day and to remove all clear eggs. The hens should not be allowed to leave the nests from the time the eggs begin to "pip" until all of the eggs are hatched. Nothing but good shaped eggs, neither too large nor too small, should be set. Thirteen eggs is the most satisfactory number.

CHICKEN LOSSES

Chicken losses, that represent nothing more than the average losses in all yards, seem to the novice very excessive. The losses would be pretty nearly unbearable, if it were not for the fact that they are no more than everyone has to submit to, and that there is always the possibility of doing better.

While it is hard for the beginner to believe it, there can be no doubt that a hatch which produces a chick for every two eggs must be considered good. We do not mean that very much better hatches than this are infrequent, because they are not, but there are surely more hatches that are less than fifty per cent of the eggs set, than there are that are more than fifty per cent.

It is very true that with good stock and good care, with reasonable freedom from accidents, more than eighty per cent of the chicks hatched should be raised to maturity. Yet experience has shown, that there are thousands of instances where eighty per cent of the chicks are lost and it is also shown, that a fifty per cent loss in the chicks, from day-old to maturity, is not prohibitive as respects the possibilities of financial success.

Then again, there is the loss that no one can prevent, that half the chicks hatched and raised will prove to be cockerels. This, of course, is not a loss in one sense, except that the main purpose in raising chicks is to produce pullets.

From the above it will be seen that there is little possibility of producing more than one layer to put in the house in the fall after culling, from every eight eggs put into the incubator. It is more than possible that ten per cent of the good pullets, actually raised and put into the house, will be removed during the first year from one cause or another. And at the end of two years there will be very few that will be worth keeping for layers for a third year.

To the beginner these figures are no doubt very discouraging. But they are not as bad as they look in print, because every person engaged in the poultry business is subject to just such average losses. Careful attention to details can reduce the losses very materially and every reduction made means additional profit.

On the basis of the above figures, it is apparent that a plant carrying one thousand layers should plan to put two thousand day-old chicks, at least, into the brooder each year. These should produce seven hundred broilers for the market and five hundred well selected pullets for the pen. If very bad luck is encountered there are always the best two year olds to keep over. If, however, especially good luck is had, the yearlings may be culled closer.

Chicks at four weeks of age should weigh from seven to nine ounces. At eight weeks should weigh sixteen to eighteen ounces; at ten weeks about two pounds. Leghorns at broiler age will weigh approximately the same as the heavier breeds.

A satisfactory way to test a lice powder, to see if it really destroys lice, is to thoroughly dust a bird and then place the bird under a box over a clean sheet of white paper. If the powder is doing its work properly, the lice will be found on the paper after a very short interval.

WHAT THE GOVERNMENT SAYS

Since writing some of the chapters of this book relating to the chances of success with poultry-keeping, especially Back Yard Poultry-keeping, and relating to location of farm, etc., we have happened to notice a discussion of the same subjects in a general way in the 1914 Yearbook of the Department of Agriculture. We believe that it is worth while to reprint here a part of what is said on the subject, as it shows that the opinions that we have expressed are shared by eminent authorities on the subject, who look at the matter from an entirely different angle than ourselves.

"The movement from city and town to farms for permanent residence has not been uniformly successful; indeed the failures have been considerable. Success in agriculture requires a great variety of knowledge and much experience. A successful farmer of the present time may need considerable knowledge of chemistry, of bacteriology, of economic entomology, of the physiology and pathology of plants and animals, of plant and animal breeding, of fungicides and insecticides, of the conservation of soil moisture, of botany, pomology, viticulture, of horticulture in general, and certainly much concerning the practical handling and marketing of his products.

"Not only must the watchfulness of the farmer be unremitting, but his essential labor is exacting, often strenuous, daily and seasonally long sustained, requiring physical endurance as well as ready adaptability. The migrant from the city has not always known of these requirements, and is then poorly prepared to meet them.

"Many city people who have gone to farms to live have failed because of inexperience and because of visionary projects. When they needed to depend on hired labor they have been unable to find laborers to employ, or, if they found them, the laborers have been too untrustworthy and ignorant, and frequently the employers have lacked experience in their management.

"Some of these farmers from the city have not been provided with sufficient capital, they have become discouraged because they have failed to get profit at once, they have had absurd expectations of profit, they have unexpectedly found a hard life, and some of them have been deceived by real estate agents into making unwise purchases of land that practically no one could have kept.

"In the movement to new land many settlers were duped by real estate agents into buying land where farming operations were carried on against great obstacles and where failure was a natural result.

"Overvaluation of land and excessive prices imposed upon it have often been a burden against which the purchaser from the city has struggled in vain. There are regions, large as well as small, throughout which agricultural land prices have risen above a fair economic level, and the situation has been disadvantageous to the new farmer with limited means. Either he has been cramped for capital or he has borrowed because of the exaggerated valuation, and he has had to work and strive, to this extent, against an artificial situation. For this reason many have failed to secure a footing as farmers, and also because of this condition many have been unable to make the venture.

LARGE PROPORTION OF SUCCESS

"On the contrary, a large proportion of the farmers from the city have successfully established themselves, although more generally in small ways than in large ones. Not all of them found poultry raising a losing venture, especially when combined with vegetables, berries and small fruits.

"In some cases, to bring in a money income, it was necessary that one member of the family should continue to work at his mechanical trade or in the mill, while the other members of the family maintained the farm. Success in the farming venture depends largely on the knowledge previously acquired, as well as on the experience that developed in the undertaking.

"The farms acquired by city families for permanent residence have usually been small ones—from 5 to 10 acres and upward to 40 or 50 acres."

INTENSIVE VS. EXTENSIVE FARMING!

"It is true that the American farmer does not produce as much per acre as the farmer in a number of civilized nations, but production per acre is not the American standard. The standard is the amount of produce for each person engaged in agriculture, and by this test the American farmer appears to be from two to six times as efficient as most of his competitors. Relatively speaking, extensive farming is still economically the sound program in our agriculture, but now it is becoming increasingly apparent that the aim must be, while maintaining supremacy in production for each person, to establish supremacy in production for each acre. The continued solution of the problem here suggested is one which now seriously engages the attention, not only of the agricultural agencies of the several states, but also of the Federal Government." (Yearbook of the Department for Agriculture. 1914.)

HOW AND WHY EGGS ARE GRADED

Believing that it will prove both of interest and of value to the novice to know what happens to the egg in the city market and what the requirements of that market are, and also what needs to be done to meet those requirements, we give the following further extract on the subject from the Yearbook of the Department of Agriculture.

"The one way to tell about the contents of an unbroken egg is to hold it before the light. Testing millions of eggs in this way has enabled the testers to tell just how each grade looks. To assist those who are not experienced candlers, the Government has printed carefully colored lithographed charts, which show the exact appearance of different grades of eggs before the light. With this chart the egg dealer, and even the housewife, is enabled to candle eggs with sufficient accuracy.

"The absolutely fresh egg held against the light shows a distinctive pinkish glow of goodness. Let that egg, however, remain out in the sun or in the summer heat for a little time, and within a day or two it begins to show "blood," a tiny series of little blood vessels forming around the embryo of the chick; or the heat may cause the yolk to go toward the top and shift easily, which characterizes it as a "light floater." Again, the yolk may mix with the white and make a "white rot," where no light at all can be seen through the egg. The egg has now reached the explosive stage, which makes it such a favorite missile of the average boy. There is, however, another type of bad egg which most people would think good for food. The yolk is a firm golden ball and the white a clear liquid. But the white has a greenish color—and the green indicates that the egg is full of bacteria—it is a "green white egg."

After the candle has told its story, the egg, if intended for long shipment or storage, must again pass examination before it has been classified fully, for an egg is no better than its shell. A perfect shell is one of the essentials of a good egg, because any crack or break in it will tend to let in all sorts of bacteria to hasten its putrefaction. The egg therefore must be graded, not only by the condition of its contents, but by cleanliness and soundness of shell as well. An egg that is so badly cracked that its contents escape is termed a "leaker." A "leaker" not only will not keep itself, but it may soil and injure a large number of eggs packed in the same case with it. They are thrown out, therefore, at every stage of handling and constitute a total loss.

"Checks" are eggs the shells of which are cracked but the membranes still intact. These too are sure to rot quickly. Even if their contents are perfectly fresh they can not be held for any long period. The egg with a dirty shell, no matter how good its contents may be does not bring a high price on the market. It is unpleasant for the housewife to handle and can not be served in the shell. Washing dirty eggs however hurts rather than helps them for the reason that any water put on an egg washes off some of the protective covering which the hen puts on the shell to make it more resistant to the entry of germs. A washed egg is shiny and smooth looking and lacks the powdery bloom of a clean fresh egg that has not been washed. One of the duties of the egg tester, therefore is to detect the egg which has been washed to escape the lower commercial grade assigned, to those with soiled shells.

The commercial egg handlers in the large cities understand fully the importance of the delicate candling tests and the careful examination and classification of eggs according to shell condition. In some cases the middlemen—largely the country merchants or egg collectors of the small railroad towns of the egg-producing districts—understand candling, but frequently conduct this operation more or less roughly and do not always grade the egg accurately. Many of them will count as nearly fresh, or "strictly fresh," eggs which under the careful candling of the cities would be put in other than the highest grades. Similarly in the other gradings their candling is not nearly so exact as the work in the great egg centers. A partial explanation for this is that the country egg collector is interested merely in getting his eggs 200 or 300 miles to the nearest wholesale egg-collection market. The big candler in the city, however, must decide what eggs are good for the fancy trade and what eggs to sell as second and third class. Moreover, in the winter the city candler must candle the eggs that have been kept for months in cold storage to satisfy the egg hunger of the great cities during the seasons when the sturdy gray hen is not laying.

Without such provision for the cold storage and canning of large supplies, a palatable boiled egg in the winter in the great cities would be a luxury entirely beyond the reach of any but the wealthy. The day when the housewife purchased her eggs from a near-by hen-keeping neighbor has passed. Few of the city dwellers ever hear a hen cackle, except at the annual poultry shows, and not one out of ten thousand could go directly to a place where she could get eggs taken fresh from the nest. The luxurious eggs which sell at from 60 cents to a dollar a dozen do come largely from the environs of the large cities, but the bulk of the egg supply travels for distances sometimes as long as 2,000 miles to reach the markets on the Altantic coast. This has transformed what once was a matter largely of personal barter between neighbors or between the country woman and the storekeeper, who supplied his own retail customers into a vast and complicated food industry employing thousands of people and many millions of capital.

Today the egg starts on its trip to the big markets in the farm wagon, and, for the woman of the farm each good egg in its shell is practically so much cash. In fact, eggs and chickens supply a large proportion of what might be called the ready spending money of the farm woman. The local merchant pays her for her eggs either in money with which she pays her doctor's or dentist's bills, or buys articles for the home or else immediately transforms the eggs into calico or shoes or groceries. The country storekeeper similarly regards the eggs as money and deposits them with the local egg collector and shipper who honors the poultry check and turns back the cash to the storekeeper. From this point of view the local egg shippers might be regarded as running an egg bank. After the egg shipper at the local station starts his eggs toward the collecting centers in the large cities, the egg then passes through practically all the selling agencies that attend the getting of any manufactured foods from the factory to the actual consumer.

The difference, however, between the vast egg industry and the big trade organizations for handling other staples, is, that while the capital involved in the egg collecting, handling, and storage business is very large, the trade is not highly centralized. The imposing total is made up of many small units—thousands of men in local towns who have invested a few hundred or a few thousand dollars in their branches of egg marketing. It is estimated that these men handle yearly a food product worth, to the consumer, $750,000,000."

The reader will notice that we have given in this chapter several extracts from the Yearbook of the Department of Agriculture. We have felt free to do so because, while it is a book of inestimable value and one that every farmer should own and read, we feel very sure that it is another one of those books issued by the Government which are usually looked upon as containing only dry statistics. Consequently we feel that this particular book has in all probability not been very widely read and we consider the few things we have quoted above as very much worth while for poultry men to know about.

As a matter of interest solely, we append herewith a chart showing the range of prices of eggs in the Boston market during the past three years. The grade of eggs used for the chart is what is known as "strictly fresh hennery" eggs as quoted daily in the Boston Globe.

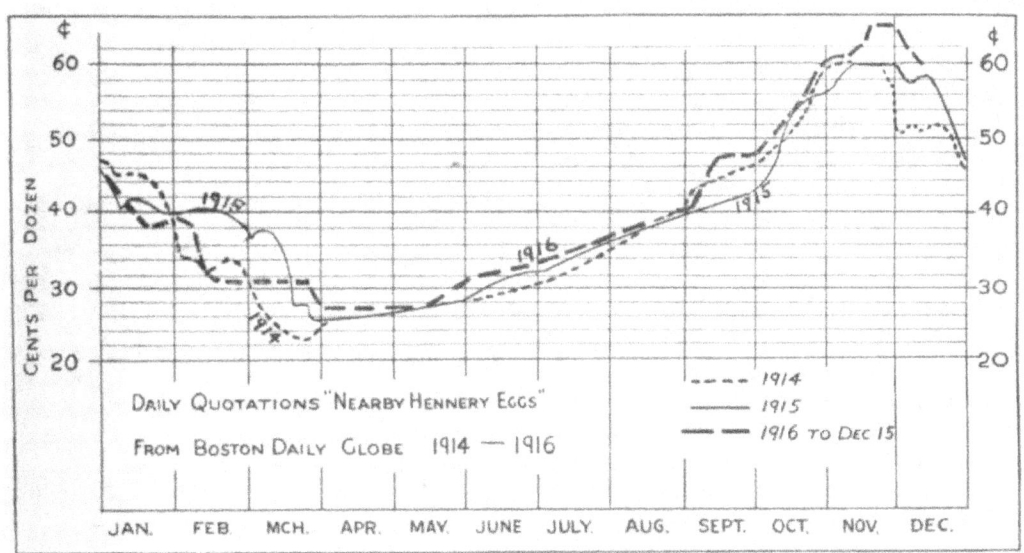

During the summer and fall months at our farms we get from three to five cents more than these quotations.

Average Price of Eggs Received by Farmers on First of Each Month by States—1914
EGGS, CENTS PER DOZEN

STATE	Jan.	Feb.	Mar.	Apr.	May	June	July	Aug.	Sept.	Oct.	Nov.	Dec.
Me.	38	32	31	22	22	22	24	26	30	33	36	45
N. H.	38	35	32	22	23	24	24	27	32	35	40	47
Vt.	39	34	31	22	20	22	20	23	27	29	34	41
Mass.	42	34	35	27	26	26	28	35	38	41	47	53
R. I.	40	38	38	26	21	25	27	30	34	39	45	55
Conn.	42	38	36	26	25	26	25	30	33	38	48	49
N. Y.	39	36	32	22	20	21	22	25	29	33	37	41
N. J.	42	37	33	26	21	23	26	28	31	35	39	45
Penn.	36	34	28	22	18	19	22	23	26	28	32	37
Del.	31	31	28	18	18	20	21	26	25	25	33	38
Md.	32	30	25	18	17	18	19	20	23	26	29	34
Va.	29	27	24	18	16	17	18	18	21	24	25	30
W. Va.	31	38	26	21	18	18	19	19	21	24	28	30
N. C.	27	23	21	17	16	17	18	18	20	23	23	26
S. C.	29	26	22	21	20	20	20	21	20	24	24	26
Ga.	29	25	22	20	18	18	18	18	20	23	24	26
Fla.	35	30	25	22	22	21	21	25	25	28	31	33
Ohio	32	30	25	17	17	18	18	19	22	25	26	32
Ind.	28	27	23	16	16	17	17	17	20	23	24	30
Ill.	29	29	25	16	16	17	16	17	19	22	23	29
Mich.	31	30	28	19	18	18	19	20	22	24	26	28
Wis.	30	29	26	17	17	17	17	18	21	23	24	27
Minn.	28	26	25	16	16	16	16	17	21	22	23	27
Iowa	27	26	22	16	16	16	16	16	20	21	21	26
Mo.	28	26	23	16	16	16	13	14	17	19	20	25
N. D.	29	26	26	16	14	14	15	15	19	21	22	26
S. D.	29	26	22	15	15	16	15	16	18	19	22	26
Neb.	28	26	22	16	15	15	15	15	17	19	21	25
Kan.	30	26	21	16	15	15	15	15	17	19	20	25
Ken.	27	25	22	16	15	16	15	15	16	19	21	27
Tenn.	26	25	20	16	15	15	15	14	16	18	20	26
Ala.	28	22	20	16	16	16	16	17	18	21	22	24
Miss.	26	22	19	17	15	16	16	16	17	21	22	23
La.	28	23	21	20	17	18	18	18	19	23	23	24
Texas	27	22	18	15	14	14	15	14	17	18	19	23
Okla.	27	24	20	15	14	14	14	13	16	17	19	22
Ark.	28	26	20	16	15	15	16	15	16	20	21	23
Mont.	40	39	33	21	18	19	22	23	26	29	35	38
Wyo.	37	34	30	21	19	21	23	23	25	27	30	34
Col.	38	36	25	21	19	19	21	20	26	29	31	33
N. Mexico	40	36	27	25	23	26	24	25	29	29	31	33
Ariz.	50	36	32	23	23	25	31	29	31	35	42	37
Utah	40	31	27	17	17	19	20	21	22	27	27	31
Nev.	52	42	32	30	29	28	30	28	37	45	45	46
Idaho	38	34	27	17	17	18	19	21	22	26	30	33
Wash.	41	34	26	19	19	20	22	25	29	33	37	40
Ore.	41	35	25	20	18	19	23	24	27	31	34	39
Cal.	45	39	25	19	21	20	24	26	29	35	41	44

CHAPTER TEN

BREEDING FANCY FEATHERED BIRDS
ESPECIALLY BARRED PLYMOUTH ROCKS

We have included this chapter on "Fancy Breeding" because we wish to give a somewhat better insight into breeding problems than can be obtained by considering the matter from a strictly utility standpoint. Many of the details, however, suggested with relation to breeding exhibition specimens are equally applicable to the breeding of utility specimens, especially in such matters as vigor, size, knowledge of the histories of the birds, etc.

It is impossible to take up the matter in this single chapter from the standpoint of the many kinds of birds that make interesting breeding, so what remarks we make with special relation to any one breed will be confined to the Barred Plymouth Rock, which is one of the most popular exhibition varieties, and which offers one of the most interesting breeding problems. The underlying principles of breeding are pretty much the same for one breed as for another, so, if we eliminate such matters as color and shape, what is said of the breeding of one variety will apply in a practical sense to any other.

MUCH PLEASURE IN FANCY BREEDING

It cannot be denied that there is a very great amount of pleasure in breeding for nicely colored, well shaped, or finely penciled birds, and that there is much satisfaction in producing a specimen that excels in these particulars. The road to success, however, in fancy breeding is very rough and narrow and hard to travel over. It means constant and uninterrupted attention to the minutest detail. It requires a thorough study of the whole history of the breed, and particularly of the history of the families of the breeds that one is working with. The short of the matter is, the fancy breeder must know what he is after and must use every means to accomplish his objects.

THE GOAL

The goal in all fancy breeding is the ideal conception of the breed as outlined and set up for us in the "American Standard of Perfection." The first step for the beginner to make in fancy breeding is to purchase a copy of this Standard and to study his breed in all its particulars. The Standard, of course, is an arbitrary one, set up by man as the ideal to work for. It is hardly likely that a bird will ever be produced to fulfill the requirements of the Standard, because if there ever should be any specimens that approached this standard, the standard would of course be raised. As a matter of fact, however, one is able to get only an occasional specimen that at all approximates the ideal.

MAKE A SURE FOUNDATION

In laying the foundation for the breeding of an exhibition line, the beginner must take pains to select a pen of birds that he has every reason to expect, in view of the ancestry of the different individuals, will mate together properly. Haphazard matings will almost always result in chicks inferior to the parents, so unless the beginner has taken intelligent care to put together birds of the same line of breeding, he is quite likely to make progress backward instead of forward. By the same line of breeding, we do not especially mean birds in as close relationship as brothers and sisters, but birds having the same general line of blood, regardless of how close the relationship may be.

A MATING THAT NICKS

We will assume that the interested reader has a pen of fairly satisfactory birds that will form a correct mating, or as it would be expressed in the parlance of the business, that will "nick" properly. He cannot expect to purchase of anyone, except at most outrageous figures a pen in which all of the individuals are of the highest possible quality. There is no one yard, in the first place, that can offer such a pen, and, if it were true that it could, the different specimens would be most likely bred from different families, which it might not do at all to breed together.

In the second place, it is unwise for the beginner to attempt to handle a pen of such quality at the price that would be involved. He could not afford to take the chance of poor results. Furthermore, his chances of success would be equally as good with brothers and sisters of these winning specimens, as with the choicest specimens themselves, as they are of the same line-breeding and the same blood line, although perhaps not so attractive to look at.

ALL IS NOT GOLD THAT GLITTERS

From First Prize Pen—Boston 1916

Unattractive hens very often produce the most attractive chickens. It is not always true that the very choicest exhibition specimens make the best breeders. Every experienced breeder knows that there are many specimens which have never been shown and that never could win in a show, that make the strongest breeders and that produce the very best exhibition specimens. This goes to furnish further proof that the breeder must know his birds thoroughly and have re'iable records to show which are his best breeders, and that he must exercise the utmost care in mating.

THE TRAP-NEST

A trap-nest is an indispensable feature in exhibition breeding, unless one can confine his matings to single males and females. By means of the trap-nest one can always know by which hen every egg is laid, and consequently which are the parents of every chick that is raised. As soon as the chicks are hatched, they should be labled with a distinctive mark, indicating their breeding. There are fifteen different combinations of holes that may be made in the webs of a chicken's feet that offer convenient means for indicating the different breedings.

THE BARRED ROCK

The Standard for Barred Rocks calls for the male and female to match in color in the show pen. Years of experience, however, has shown that it is impractical to produce Barred Rock males and females from a single mating, that will meet this requirement. Consequently, a system known as double mating has been adopted, which consists of one mating to produce males of a standard color, called a cockerel mating, and another mating to produce females of a standard color, called a pullet mating. If males and females of standard color in Barred Rocks were mated together, the result would be males that were too light in color and females that were too dark in color for exhibition purposes, except in very few instances. It is probable that there would be some few specimens produced of both males and females of the proper shade. The tendency in a standard, or single mating, however, is for the males to come lighter and the females darker than the parent stock.

COCKEREL MATING
THE MALE

In this mating a male, cockerel or cock, should be selected, that is as near standard in shape and color as it is possible to have. The bird must have strong color, with barring that is narrow, and straight cut across the feather, the dark bar being a bluish-black and distinct for the entire length of the feather. The head and comb should approximate the standard as nearly as possible, and special care should be taken that the bird selected to head the pen is thoroughly healthy, active and vigorous. Vigor should be a watchword always. One must remember that the male is half the pen.

THE COCKEREL FEMALES

The females in the mating should be two or three shades darker than the standard and be rich in many generations of pure cockerel blood. They should have neat heads, evenly serrated combs, full breasts, medium length backs, well barred tails, and wings, and strong narrow barring with good undercolor and good size. The dark bars on the feather should be twice as wide as the light bars.

It is suicide to breed from an undersized female in this mating, for males, to be attractive and useful, must be good size. Breeding from small females will produce birds with small bones, and they would always be undersized. It is desirable also, that the feathers, as far as possible, each end with a dark tip.

While it is a great temptation to select for this mating females that are very handsome with sharp snappy barrings, we find that the less attractive females are usually the best producers of exhibition cockerels, and should they have a little brown or muddy look, they are probably even more valuable breeders. Be careful in selecting the females that they have very dark barred hackles, as females with light colored hackles are very sure to produce cockerels uneven in color.

Color Special Male, Boston

PULLET MATING
THE FEMALES

The females in a pullet mating should be as near the standard shape and color as possible. They should be birds of good size, having full breasts, very evenly serrated low combs, red eyes, backs of medium length and good width, and tails well opened. The barrings should be sharp and snappy, straight and clean cut across the feather. The dark bars should be a bluish black color. The dark and light bars should be of equal width and should show clearly and distinctly the entire length of the feather. Particular care should be taken that the barring is good on wings and tails, and that all feathers end with a dark tip.

THE PULLET MALE

The male should be what is known as a pullet breeder, a bird bred from a similar pullet mating. It should be of good size, stand well on its legs and be a little more rangy than is called for in the standard shape. His color should be two or three shades lighter than the standard, the barring being straight cut and clear and distinct in all sections. The dark bars should be very narrow.

Where females are used that are strong in under color, we would select a male that is not too strong in this particular, one that shows under color rather faintly. If both males and females are used that have strong tone in the under color, the surface color on the female chicks produced is likely not to be clear and clean. The pullet males should be especially strong in legs and beak; these should be a nice, brilliant yellow. The combs should be evenly serrated and, furthermore, should be low. As in all breeding birds, pay special attention to selecting birds of good vigor.

THE LEFT-OVERS

Every breeder, after he has made up the matings that are possible from the few very choice specimens among his stock, has quantities of birds left over of varying shades of color, shape and size. Some may have excellent shape, but be deficient in barring; others may have excellent barring and shape, but be too light or too dark. Any breeder usually finds that the number of choice matings that he can make from his birds, no matter how many birds he may have, is very much limited, because so few birds anywhere near approach the ideal.

VALUABLE TO USE

It is not true however that it is impractical to use these birds that are more or less deficient in one respect or another. A little careful study and experimenting will permit the breeder to mate these females so that they will produce standard shape and color. We say females because it is rarely worth while to use a male that is deficient unless he may be a bird that is strong in the respects that certain females that it is desired to use are weak.

Good results can be obtained by mating to females too light in color, but otherwise good quality, a pullet bred male that is dark in color to about the same degree that the females are light. Vice-versa, a light male may be used with females that are too dark. We have had matings from which we have had excellent results, where the male was so light that it showed practically no regular barrings. Extreme measures of this sort can be taken, but only when it is known that the bird used was produced from a strong exhibition line.

White Plymouth Male Heading First Prize Pen, Boston 1916

JUST AN INSIGHT

The above remarks, of course, cover the subject of fancy breeding in only a very superficial way, but they may serve to give the novice some insight into the attention necessary to minute details, a general idea of the kind of methods used, and may impress upon him the necessity in such breeding of using only stock that has been bred carefully and intelligently and whose history is known. If this chapter has accomplished this it has served its purpose.

NOTES ON BREEDING

A bird that may have an injured leg or comb, may be as valuable as a breeder as some other birds that look more perfect. A crooked breast bone should not cause you to discard a bird that is otherwise good, if you know that the deformity was caused by accident, and is not due to constitutional weakness.

Be sure to have any birds, that you are to display in the show room, well tamed and accustomed to handling, as many judges do not take kindly to a bird that objects to handling or that is ugly.

A female that is off color, but otherwise well marked, and that contains the blood of a long line of distinguished ancestors is quite likely to prove a good breeder.

It is more advisable to introduce new blood through a female than through a male. If the union should not prove what was expected you would not lose the benefit of the whole pen.

March, April and May are the best months to hatch for December and January shows.

If you get ten high scoring birds from a mating of ten birds, you may feel that you have been favored with at least average results.

EACH HEN REQUIRES

55 pounds scratch grain per year.
30 pounds mash per year.
15 pounds other food per year.
50 to 100 square feet yard space.
8 to 10 inches roosting space according to breed.
4½ to 6 square feet floor space in small houses.
3½ to 4 square feet floor space in very large houses.
One nest to every four birds in small pens.
One nest to every five birds in larger pens.
Water, grit, charcoal, oyster shells and dust bath all the time.

SEED PER ACRE

Alfalfa	20-25	pounds
Beet	3-7	"
Clover	16-18	"
Corn (Hills)	6-12	quarts
Corn (Drills)	2½	bushels
Barley	2	"
Buckwheat	½	"
Oats	3	"
Rye	1½	"
Wheat	2	"
Rape	5-8	pounds

Appendix

Handy Recipes

BORDEAUX MIXTURE

Copper Sulphate (blue vitriol) 4 pounds
Quicklime (not air slaked) 4 pounds
(Of dry air slaked lime or hydrate of lime one-fourth more.)
Water to make 50 gallons.

Dissolve the copper sulphate in about two gallons of hot water, contained in a wooden vessel, or even better by suspending the sulphate contained in a cheese cloth sack, in a large bucketful of cold water. With the cold water and cheese cloth bag a longer time is required. Pour the sulphate solution into the barrel or tank used for spraying, and fill one-third to one-half full of water. Slake the lime by addition of a small quantity of water, and when slaked cover freely with water and stir. Pour the milk of lime thus made into the copper sulphate, straining it through a brass wire strainer of about 30 meshes to the inch. Pour more water over the remaining lime, stir and pour into the other; repeat this operation until all the lime but stone lumps or sand is taken up in the milk of lime. Now add water to make fifty gallons in the tank. After thorough agitation the mixture is ready to apply. The mixture must be made fresh before using, and any left over for a time should be thrown out or fresh lime added.

USES—For fungus pests,—leaf spot,—rust,—blight,—mildew, etc. on apple, cucumber, melon, sqaush, tomato, etc.

KEROSENE EMULSION

Laundry soap (chipped) ½ pound.
Kerosene (coal oil) 2 gallons.
Water (preferably soft and free from dirt particles) 1 gallon.

Dissolve the soap in the full amount of water and when this solution is boiling hot remove from the fire and add the kerosene. Stir the mixture violently by driving it through a force pump back into the vessel until it becomes a creamy mass that will not separate. This requires usually from five to ten minutes. For use, dilute one part of the emulsion with 8 or 10 parts of water for scale insects and hard-bodied insects like the cinch bug. For soft-bodied insects, such as plant lice, lice on animals, etc., use one part emulsion to 15 or 20 parts of water. The stock emulsion will keep good for months if kept in air-tight vessels.

Kerosene emulsion kills by contact, and therefore the application should be very thorough. It may be used against a great many different pests, but is sepecially valuable for destroying those with sucking mouth-parts for they cannot be killed with arsenical poisons.

CAUTION—Only the dilute emulsion, 1 part emulsion to 15 or 20 of water should be used when the trees are in leaf, and in all cases it should be kept thoroughly stirred; also apply only on bright, sunshiny days, otherwise the foliage or even the twigs will be injured.

USES—For Elm scale,—Oyster shell scale,—Rose scale,—Mite,—Red Spider and other insects.

ARSENITE OF LIME

White Arsenic, 1 pound
Lime, 2 pounds
Water, 3 gallons

Boil together for full 40 minutes after the boiling point is reached. As precaution against danger of burning of foliage, slack an additional pound of lime, add to it three or four gallons of water, and add to the boiled mixture. Strain and dilute to form 200 to 250 gallons for hardy vegetation such as potatoes. Do not use at all on stone fruits or on cucurbits. Dilute to 300 to 400 gallons for tender vegetation.

USES—For Elm Beetle,—Plant Lice,—Cut-worm,—Codlin Moth,—Willow Worm.

WHALE-OIL SOAP SOLUTION

Whale-oil soap, 2 pounds
Water, (hot) 1 gallon
Dissolve the soap in one gallon hot water. Dilute four times before spraying.

USES—Mealy Bug,—Roseleaf Hopper,—Scale Insects,—(A weaker solution will kill plant lice.)

COPPER SULPHATE SOLUTION

Copper sulphate, 4 pounds
Water to make 50 gallons

Dissolve the sulphate as directed in Bordeaux mixture.
CAUTION.—This solution will injure foliage. It can be used only before the buds open.

USES—For small fruits.

GLYCERINE CEMENT

Litharge and Portland cement (equal parts)
Glycerine

Litharge and Portland cement, equal parts, mixed with Glycerine to the consistency of dough.

USES—For stopping leaks in tanks, in fact, for closing cracks and stopping leaks in almost anything. Where only small amount is to be used, the cement may be omitted.

CEMENT MORTAR

Portland Cement, 1 barrel
Sand, 4 barrels
Lime putty, 2 pails

USES—For laying brick or stone.

"LEAN" CONCRETE

Portland cement, 1 barrel
Sand, 4 barrels
Gravel or broken stone, 8 barrels.—A 1:4:8: mixture—

USES—For unimportant work, Hen house floors, etc.

ORDINARY MIXTURE

Portland cement, 1 barrel
Sand, 3 barrels
Gravel or broken stone, 6 barrels—1:3:6: mixture—

USES—For retaining walls, piers, etc.

SIDEWALK MIXTURE

Portland cement, 1 barrel
Sand, 2½ barrels
Gravel, or broken stone, 5 barrels—1:2½:5- mixture—

USES—For sidewalks, floors, walls, arches, fenceposts, etc.

RICH MIXTURE

Portland cement, 1 barrel
Sand, 2 barrels
Gravel, or broken stone, 4 barrels—1:2:4: mixture—

USES—For engine foundations, water tanks, floor beams, etc.

AMOUNT OF CEMENT NEEDED FOR ONE CUBIC YARD OF EACH ABOVE FORMULAE

1:4:8 Mixture requires   0.9 barrels cement
1:3:6 Mixture requires   1.1 barrels cement
1:2:5 Mixture requires   1.3 barrels cement
1:2:4 Mixture requires   1.6 barrels cement

LAND MEASURE

144 square inches equal 1 square foot
9 square feet equal 1 square yard
30¼ square yards equal 1 square rod
40 square rods equal 1 rood
4 roods equal 1 acre 43,560 square feet

One section is one mile square, or 640 acres.
A square acre is 208.71 feet on each side.

METRIC MEASURE

One metre is approximately 39.38 inches.
One kilogram is approximately 2.2 pounds.
One liter is approximately .26 gallons.

WEIGHTS OF MATERIALS

Wheat, 60 pounds per bushel
Corn, 56-58 pounds per bushel
Oats, 32 pounds per bushel
Barley, 48 pounds per bushel
Hard coal, 50-55 pounds per cubic foot.
Soft coal, 40-50 pounds per cubic foot.
Cast iron weighs, a quarter pound to the cubic inch
Common earth, 75-90 pounds per cubic foot
Sand, 115-130 pounds per cubic foot
Granite weighs, 165-175 pounds per cubic foot.
Water weighs, 62.4 pounds per cubic foot.
Ice weighs about, 57 pounds per cubic foot.

TABLE OF SEED VALUE

	Seeds for 100 foot row	Plants from one ounce of seeds	Days to mature
Asparagus	100 plants		1100
Beans	½ to 1 quart		45-100
Beet	2 oz.		60-90
Cabbage	—	2,000	100-150
Carrot	1 oz.		90-125
Celery	2 oz.		130
Chard	2 oz.		75
Corn	1 quart		75-100
Cucumber	—	75 hills	75
Lettuce	½ oz.	1,000	75-100
Onion	1 oz.		150
Parsnip	½ oz.		125
Pea	1 quart		60-75
Radish	1 oz.		30-50
Spinach	¾ oz.		60-75
Tomato	—	1800	140
Turnip	1 oz.		60-75

LIME METHOD

Slack four pounds of quick lime in a small quantity of water, add four gallons of water, stir the mixture several times for two or three days, then let it settle. Use the clear liquid in the same manner that the water glass mixture described above is used.

WHITEWASH

Place ten pounds of lime in a pail or cask and pour two gallons of water over it; after fifteen minutes it may need a little additional water. Let stand an hour or more and reduce to consistency of cream.

If used in a spray, it should be reduced to about one tenth original strength and should be strained through a fine cheese cloth.

PRESERVED EGGS
WATER GLASS METHOD

In the proportion of ten quarts of boiled water to one quart of water glass, thoroughly mixed, fill a clean jar, either of wood, glass or earthenware, to whatever depth is required to cover the number of eggs to be "laid down." The eggs should be placed in the mixture small end down. Do not stir the mixture after the eggs have been put in, but water should be added as the same evaporates. Keep in a cool cellar with the jar lightly covered. Infertile eggs laid in April or May are the best for the purpose.

WEIGHTS OF THE DIFFERENT KINDS OF ROOFING

	Lbs. per Sq. Ft
Cast Iron Plates	15
Copper	.8 to 1.25
Felt and Asphalt	1
Felt and Gravel	8 to 10
Iron, Corrugated	1 to 3.75
Corrugated sheets, unboarded	8
Iron Galvanized, flat	1 to 3.50
Sheathing, pine 1 inch thick, yellow	3 to 4
Shingles on lathes	10
Spruce, 1 inch thick	2
Spruce, if plastered below rafters	12
Sheathing, 1 inch chestnut or maple	4
Slate on lathes	13
Slate on boards 1¼ inch thick	16
Sheet iron $\frac{1}{16}$ inch thick	3
Sheet iron and lathes	5
Skylights, glass $\frac{3}{16}$ inch to ½ inch	2.50 to 7
Sheet Lead	5 to 8
Tin	.7 to 1.25
Tiles, flat	15 to 20
Tiles, grooved and fillets	7 to 10
Tiles, pan	10
Zinc	1 to 2

For spans over 75 feet add 4 pounds, per square foot to the above loads.

Snow weighs 5 pounds to 12 pounds per cubic foot depending upon the humidity of the atmosphere; 1 cubic foot of snow compacted by rain weighs 15 pounds to 50 pounds. It is customary to add 30 pounds per square foot to the above for snow and wind when separate calculations are not made.

The weight of any load upon a roof is taken as uniformly distributed over the surface of the roof. The total weight on each pair of rafters, couple or truss, is equal to the sum of the weights of the truss itself, and as much of the roof as is carried between two trusses.

ADVERTISING SECTION.

As stated at the outset, in writing this book we have tried to discuss all subjects with as little bias as possible. We acknowledge that it is impossible to speak entirely without bias regarding such equipments and materials as we have used over a long period of years without having had occasion to adversely criticise them in a single instance. On the other hand we doubt if even the makers of other equipments or materials will consider that we have been any more than just in speaking favorably of these things.

This much the reader may be sure of, we have suggested nothing that we do not thoroughly believe in. No amount of money could tempt us to recommend anything that we do not know to be of real value.

In the following advertising section we have allowed nothing to appear that has not been thoroughly tested by us and found in our own experience to be as represented. The materials advertised we are sure can be used on every poultry farm with profit.

The four poultry papers advertised should be read by all interested in poultry. If you have anything to sell our experience tells us they will bring the best and the quickest results.

We have also ventured to advertise our own day-old chicks because we believe we may safely maintain that they are at least equal to the best.

PITTSFIELD POULTRY FARMS COMPANY, Publishers

Pittsfield Poultry Farms Chicks are practically all raised on WIRTHMORE Chick Feed because it's best.

CHAS. M. COX CO. BOSTON, MASS.

The Truth About Poultry

Get the Facts by Reading

The One-Man Poultry Plant

Successful Methods of Men on Farms or Small Acreage.
Complete in twelve parts, printed in one volume.
By DR. N. W. SANBORN

REAL work, with real poultry, on a real New England Farm. This is a simple story of what has been done by a man, at forty-five years of age, town-bred and city-educated getting out of practice of medicine, buying a small farm in the hill country, and making a success of the venture. Not only is the rearing of chicks and the management of adult fowl completely covered, but the interesting side issues of fruit growing, grain-raising and the production of milk, that cannot be escaped on a real farm. You get rugged facts — rarely found in print. The truth about poultry as found in actual life on a one man poultry farm.

You Can Do the same — Book Tells How

Our Special Offer — The One-Man Poultry Plant, in twelve parts (book form), and the American Poultry Advocate, one year, for only 50 cts., book and Advocate, three years, for only $1.00, if order is sent at once.

Our paper is handsomely illustrated, practical, progressive and up-to-date on poultry matters. Established 1892. 44 to 132 pages monthly. 50 cents a year, 3 months' trial 10 cents. Sample copy free. Catalogue of poultry literature free. Address

AMERICAN POULTRY ADVOCATE.   622 Hodgkins Blk., Syracuse, N. Y.

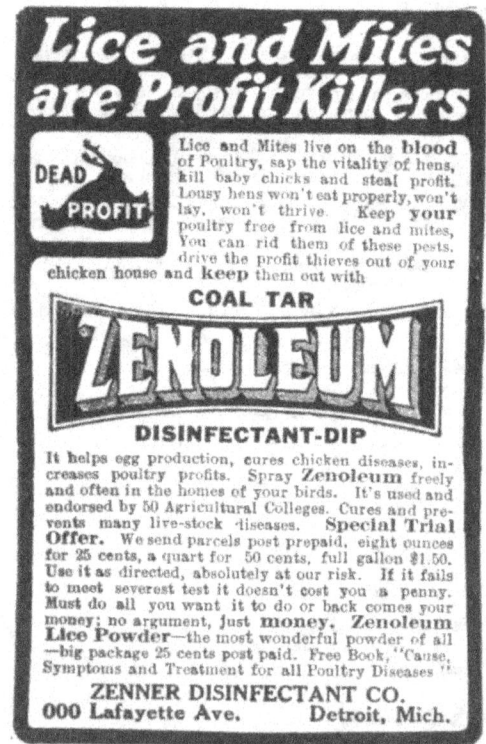

PITTSFIELD POULTRY FARMS

We are using Zenoleum in increasing quantity each year, and for precisely the same purpose for which we started to use it, namely, incubation. The fact that we are still using it for dipping eggs, after several years of thorough trial of it, would seem to be convincing evidence that we consider it very much worth while. We are now setting 185,000 eggs at a setting and ZENOLEUM is used in every hatch, just as it used to be when we only set 6,000 at a time.

F. W. BRIGGS, Manager.

FANCY FEATHERED DAY-OLD CHICKS

OUR "Gentleman's Fancy Day-Old Chicks" might be said to be in a class between our Utility and our Exhibition classes. They have the feather to please the most fastidious and the show qualities to win in the smaller shows; they are of course, somewhat more expensive than Utility chicks, but no more so than some Utility chicks offered by small breeders.

We have Gentleman's Fancy Chicks only in Barred Rocks, White Rocks and White Leghorns. If you want to get any of this class of chicks you should send us your order in advance as we have only a very limited supply. We can not make shipments of over 500 of these chicks any one week while we can ship over 30,000 Utility chicks within that time.

PITTSFIELD POULTRY FARMS CO.
Holliston, Massachusetts.

There's Only One

POULTRY MAGAZINE that regularly presents original monthly articles by the best known experts in breeding, teaching and judging, including Messrs Chas. D. Cleveland, Prof. H. R. Lewis, T. F. McGrew, W. C. Thompson, Theo. Hewes, Mrs. I. F. Rice, etc. Seasonable articles will insure **Your Poultry Success.** "Everybody reads Everybodys" 68 to 150 pages monthly of the best matter in educational articles and reliable advertising. Send in your subscription now, to-day, don't miss a copy at fifty cents the year. Three years for **One Dollar.**

HENRY P. SCHWAB, Editor and Manager.

EVERYBODY'S POULTRY MAGAZINE
HANOVER, PA.

Are you going to raise CHICKS

8 weeks' start on Milk Mash. Costs less to feed than skim milk

IF you will just give them the right start they will give you a profitable finish.

Blatchford's Milk Mash is a dry milk equal —just the thing for their tender digestive organs, because it's made expressly for them.

Bowel trouble cannot exist with your chicks when you feed Blatchford's Milk Mash

A Postal Card will bring you further particulars.

Blatchford Calf Meal Factory.

In business over one hundred and fifteen years.
Over thirty-five years in the United States

Waukegan :: :: Illinois

World's Largest Manufacturers of Poultry Supplies

The O. B. Andrews Barred Rocks received the greatest number of points at the recent PALACE SHOW. Every article manufactured has been thoroughly tested in connection with the raising of these wonderful birds.

SEND FOR CATALOGUE.

O. B. ANDREWS CO., Chattanooga, Tenn.
Dept. PP

"Your Valuable Poultry Journal is Well Worth MANY TIMES the Subscription Price."

➤ Thus speaks Lewis C. Ayres, of Bryceton, Sask., Canada

THE true value of The Poultry Item lies in its Open Editorial Policy. It invites its subscribers to contribute stories of their own individual success. These are extremely interesting to the new beginner. He reads what obstacles the other fellows overcome and avoids them in his own experience.

The Poultry Item's liberal policy in printing Full Show Awards helps you keep in touch with the winnings of your special breed, and its Expert Advice Department brings the poultry doctor right to your elbow.

Every month Poultry Experts review the Industry, giving you the benefit of their experience.

Illustrations: The best to be had.

Published in the Heart of the Poultry Buying Industry, the Poultry Item's Display and Classified Advertising columns offer a world of opportunity for the quick selling of Stock, Eggs or Baby Chicks. Write for advertising Rates.

Subscription Price, 50c. per year, 3 years for $1.00. Canada, 75c. Foreign, $1.00.

THE POULTRY ITEM, 5 Maple Avenue, Sellersville, Pa.

LEARN TO BREED WINNERS
These Books Tell You How To Do It

The Wyandottes, all varieties *The Rhode Island Reds*
The Plymouth Rocks *The Leghorns*
The Orpingtons *The Campines*
Ducks and Geese *The Asiatics* *Turkeys*

Numerous Special Illustrations by
FRANKLANE L. SEWELL and ARTHUR O. SCHILLING.

Size of each book 9 x 12 inches. Range in pages from 80 to 160. Write for description and prices. 50 cents, 75 cents and $1.00 each. Special offer with a year's subscription to A. P. W. Address

AMERICAN POULTRY WORLD

85 Dewey Avenue   Dept. D   *Buffalo, N. Y.*

"PITTSBURGH PERFECT" FENCES

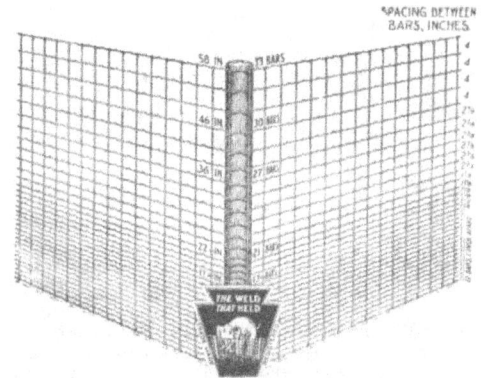

"Pittsburgh Perfect" Chicken and Rabbit Fence is the closest mesh fence of its kind in the world. For 17 inches from the ground the wires are only 1 inch apart so that your smallest chicks can't get through. All wires made of full gauge heavily and evenly coated with pure zinc galvanizing and ELECTRICALLY WELDED together at every crossing point.

Easily erected on steel, concrete or wooden posts. Low in price. Will last for many years. Made in 5 heights with stay wires either 6 or 4 inches apart.

Sold by Dealers Everywhere.   Write for Free Catalogue.

PITTSBURGH STEEL COMPANY
PITTSBURGH, PA.

New York,   Chicago,   San Francisco,   Duluth,   Memphis,   Dallas.

Manufacturers of "PITTSBURGH PERFECT" brand of Nails and Wire products.

HERE!
Write Your Own "Money-Back" Guarantee

You know what a good brooder ought to be and to do; you know how to express that in writing. Just sit down and write it out, send it to us with an order for our brooder, and we will sign the guarantee and send you the brooder on a thirty day's trial. If it doesn't come up to *your* guarantee, send it back and we will refund the money without a question.

STANDARD COLONY BROODER
PATENTED

is the greatest, most practical coal-burning brooder ever made. Self-feeding, self-regulating, everlasting. Broods 100 to 1000 chicks at a guaranteed cost of less than 6 cents a day. It will do anything any other brooder will do, regardless of price, and do it better.

BEWARE OF IMITATIONS

Book of Proof—Free. Write for it or ask your dealer.

The Buckeye Incubator Company
281 Euclid Ave.   Springfield, Ohio

SPECIFICATIONS

Solid cast iron stove. 52 inch galvanized hover. Two double-disc thermostats, tandem hitched. Rocker furnace grates, self-cleaning and anti-clog. Check valve hung on knife edge bearings. Gas proof—fire proof—fool proof. Guaranteed to burn more than 24 hours in any temperature with one coaling. Capacity, up to 1000 chicks.

Agents Wanted Some good territory still open. An attractive proposition for the right man or firm.

$16.00

A Little Higher in the West on Account of Freight.

Used Exclusively by Pittsfield Poultry Farms Co.   Ask them what they think about the Standard Brooder

CHIC  CHUK
TRADE MARK REGISTERED

THE CONCENTRATED POULTRY FOOD

50% Protein, 30% Bone, 2% Fat, 2% Salt.

5 & 10 lb. Cartons
30 & 100 lb. Bags

READ THIS LETTER

RUSSIA CEMENT COMPANY
    Gloucester, Mass.

    We have placed our order with you for our 1917 requirements of your fish meal "Chic-Chuk" and we are especially anxious that you make prompt shipments.

    We, like all other poultry raisers, have to use a food rich in Protein to balance the rations given our birds and we recognize in your fish meal the most wholesome product we have seen for this purpose.

    Beef Scrap, which is made largely from diseased animals, is the one product we have been obliged to use that we could not be sure was wholesome.

    Your fish meal, which we have found is made from fresh waste from the boneless fish packing companies, we think is an especially attractive food and is put up in a very convenient form.

    PITTSFIELD POULTRY FARMS CO.
Dec. 1, 1916.   (Signed) HOWARD GILMORE, Pres.

Very Valuable For Growing Chicks

CHIC-CHUK is a finely ground and sifted fish meal made only from wholesome food fish—pure, sweet, Cod, Haddock and Pollock. Its high-protein content combined with vital mineral foods, means quick growth and strong, healthy chicks. Its purity makes it absolutely safe to use for chicks of all ages, as a part of grain mixtures and moist or dry mashes.

CHIC-CHUK makes pullets and hens lay plenty of good, large eggs and will increase greatly your poultry profits.

Write for our *free* booklet "Practical Talks on Poultry Feeding." It contains the very latest information on the subject, by a practical poultryman and a well-known authority.

Address
Russia Cement Co. Gloucester Mass.
Manufacturers of Le Page's Glue.

PITTSFIELD PRODUCTS

Day-Old Chicks. (Utility)

FIVE POPULAR BREEDS
- Barred Plymouth Rocks
- White Plymouth Rocks
- S. C. White Leghorns
- S. C. Rhode Island Reds
- White Wyandotts

Write for latest catalog and prices. Annual reduced prices begin early in May.

Day-Old Chicks (Gentleman's Fancy.)
Barred and White Rocks.   S. C. White Leghorns.

Hatching Eggs.
Utility, Gentleman's Fancy and Exhibition.

Breeding Stock.
Gentleman's Fancy and Exhibition.

We ship any number of day-old chicks from 25 to 10,000 at one shipment and guarantee arrival in good condition.

Advance bookings without deposit. No money down policy. Just say, How Many, When and What Breed.

Barometer of Growth.
- 10000 Egg Capacity up to 1910
- 24000 Egg Capacity 1911
- 36000 Egg Capacity 1912
- 48000 Egg Capacity 1913
- 117000 Capacity 1914
- 187000 Capacity 1916

Chart shows number of eggs used at one setting each year for past seven years. We set nine times each year.

Holliston Incubator Capacity 170000 Eggs

Pittsfield Poultry Farms Co.

HOLLISTON,   ::   MASSACHUSETTS

BLUE PRINTS.

The following prints are available to readers of this book. Each print is 14" x 18". Price 50 cents each postpaid or $5.00 for full set of 13 prints. See Chapter IV.

No. S-48. A 4 x 8 Colony House. The simplest and cheapest colony house to range 40 chicks to maturity. For most farms we would recommend the next print as being more successful.

No. S-68. A Colony House just enough larger than the 4 x 8 to protect the chicks better but not enough larger to add materially to the expense.

No. S-810. An 8 x 10 Colony House that will range 100 chicks to maturity. It will also brood 250 chicks or house 12 to 15 layers. It is the largest house that can be easily moved anywhere on the farm and is the smallest house that can be used economically as a brooder house.

No. 1414. The best size **Colony House** (14 x 14) for brooding with **Coal Burning Colony Brooders.** It will brood 300 to 500 chicks. It is on skids and can be moved short distances to give chicks new land. It will range 150 chicks to maturity or will comfortably house 35 to 50 layers.

No. 1216. A Laying Pen 12 x 16. May be used as a single pen or as a section of a long laying house. Each pen will house 35 to 50 layers.

No. 140. A Long Permanent Brooder House for use of Coal Burning Colony Brooders. The house consists of a series of pens each 14 x 14. All details are worked out and explained. It has ample ventilation, access to yards on both sides of house and every facility for economical handling of chicks.

No. S-100. Copyrighted **Feed-Chart.** Printed on cloth. This is designed to go in the feed room as a **Reference Chart** to refer to when deciding what method of feeding to pursue to meet certain conditions, It tells at a glance what each chick requires under various conditions. whether it is one day old or two years old.

No. 101 Nine New Notions. Two styles of portable yards. An improved wall nest. A stand to hold grain bags. Pittsfield Drop Curtain. Two Feed Hoppers and a water Drip System.

No. 6060. (3 prints) **Pittsfield Twentieth Century Laying House.** A laying house 60 x 60 for 1000 layers. It combines economy of construction and economy in operation better than any laying house ever devised. See description in Chapter IV.

No. 4248. Twentieth Century Jr. Laying House. Same as above only smaller (42 x 48) for 500 layers. Also in three parts. $1.50 complete.

Any of the above prints may be returned to us Postpaid if not found as **expected** and we will remit the price paid.

PITTSFIELD POULTRY FARMS COMPANY
Holliston, Massachusetts.

www.ingramcontent.com/pod-product-compliance
Lightning Source LLC
Chambersburg PA
CBHW062357220526
45472CB00008B/1838